UP-TO-S[...]

BOOK 1

STORY PROBLEMS

by Michael S. Silverstone
Thomas H. Hatch

UP-TO-SPEED MATH

Number Sense
Operations
Working with Fractions, Decimals, and Percents
Geometry and Measurement
Patterns, Functions, and Algebra
Data Analysis and Reasoning

Math Standards Review 1
Math Standards Review 2

Story Problems Book 1
Story Problems Book 2
Story Problems Book 3

created by **Kent Publishing Services, Inc.**
designed by **Signature Design Group, Inc.**

SADDLEBACK PUBLISHING, INC.
Three Watson
Irvine, CA 92618-2767

E-mail: info@sdlback.com
Website: www.sdlback.com

Copyright © 2003 by Saddleback Publishing, Inc. All rights reserved. No part of this book may be reproduced in any form or by any means, electronic or mechanical, including photocopying, recording, or by any information storage and retrieval system, without the written permission of the publisher.

ISBN 1-56254-521-3

Printed in the United States of America

TABLE OF CONTENTS

PART I PROCESS AND SKILLS

UNIT 1   GETTING STARTED

Lesson 1   Step 1 – Gather Information – Understanding the Question 6
Lesson 2   Step 1 – Gather Information – Using Clues 8
Lesson 3   Step 1 – Gather Information – Recognizing Important Information . 10
Lesson 4   Step 2 – Plan – Choosing an Operation 12
Lesson 5   Step 2 – Plan – Problems With More Than One Answer 14
Lesson 6   Step 2 – Plan – Writing an Equation 16

UNIT 2   FINDING SOLUTIONS

Lesson 1   Review 18
Lesson 2   Step 3 – Calculate – Choosing a Method 20
Lesson 3   Step 4 – Calculate – Making a Table 22
Lesson 4   Step 4 – Check – Does Your Answer Make Sense to You? 24
Lesson 5   Step 4 – Check – Using Inverse Operations to Check Your Answers . 26
Lesson 6   Sharing Your Thinking 28

PART II PRACTICE WITH STRATEGIES

UNIT 3   MULTIPLYING AND DIVIDING WHOLE NUMBERS

Lesson 1   Skill Tune-up 30
Lesson 2   Draw a Picture 32
Lesson 3   Make a List 34
Lesson 4   Work Backwards 36
Lesson 5   Choose a Strategy 38
Lesson 6   Multi-step Story Problems 40

UNIT 4   PATTERNS AND RELATIONSHIPS

Lesson 1   Skill Tune-up 42
Lesson 2   Describe the Changes 44
Lesson 3   Make a Prediction Chart 46
Lesson 4   Use an Analogy 48
Lesson 5   Choose a Strategy 50
Lesson 6   Multi-step Story Problems 52

TABLE OF CONTENTS

UNIT 5   **TIME AND DISTANCE**
Lesson 1   Skill Tune-up 54
Lesson 2   Make a Time and Distance Graph 56
Lesson 3   Draw a Clock Face 58
Lesson 4   Draw a Map 60
Lesson 5   Choose a Strategy 62
Lesson 6   Multi-step Story Problems 64

UNIT 6   **MONEY AND PERCENTS**
Lesson 1   Skill Tune-up 66
Lesson 2   Make a Tally Chart 68
Lesson 3   Use Percentages or Proportions 70
Lesson 4   Choose a Strategy 72
Lesson 5   Multi-step Story Problems 74

UNIT 7   **ADDING AND SUBTRACTING FRACTIONS**
Lesson 1   Skill Tune-up 76
Lesson 2   Make a Circle Graph 78
Lesson 3   Finding Equivalent Fractions 80
Lesson 4   Choose a Strategy 82
Lesson 5   Multi-step Story Problems 84

UNIT 8   **GEOMETRY AND MEASUREMENT**
Lesson 1   Skill Tune-up 86
Lesson 2   Draw a Diagram 88
Lesson 3   Use a Formula 90
Lesson 4   Choose a Strategy 92
Lesson 5   Multi-step Story Problems 94

TO THE STUDENT

This book contains a sequence of lessons designed to help you master the art of solving word and story problems. The lessons are organized into two sections. In Part 1, you'll learn the four-step process for problem solving. The logo at the upper left corner of each lesson will remind you to keep thinking about this process as you work.

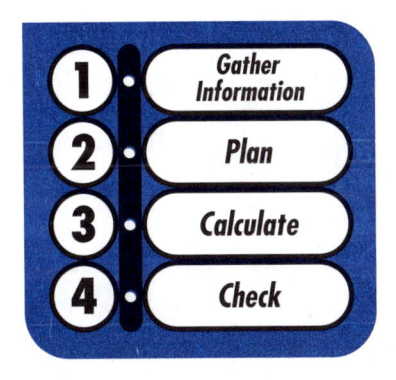

1. Gather Information
2. Plan
3. Calculate
4. Check

In the lessons in Part 2, you'll learn and practice specific math skills and problem solving strategies, such as:

- Act it out
- Draw a picture or diagram
- Guess and check
- Look for a pattern
- Make a graph
- Make a list
- Make a table
- Use proportions
- Work a simpler problem
- Work backwards

Story Problems lessons are two page spreads with instruction and worked out examples in the upper part of the left hand page. The exercises that follow the instruction provide guided and independent practice. So, get up-to-speed, and look forward to becoming a master problem solver.

UNIT 1 LESSON 1

UNDERSTANDING THE QUESTION

1. Gather Information
2. Plan
3. Calculate
4. Check

When you look at a story problem, sometimes you have to read it carefully to make sure you understand what the question is asking.

Corey wants to build a model train that is 49 cars long. He already has 27. How many more does he need?

We know Corey wants to build a long train. He needs more, though. We have to figure out how many more 49 is than 27. We can use subtraction to find out.

From the length of the train he wants to build: 49 cars

Subtract how many he already has: -27 cars

The difference is how many more he needs: 22 cars

A Which best describes what the question is asking?

1. Palace, Sabrina, and Leon want to buy a pizza that costs $8.95. Palace has $4.25, Sabrina has $2.00, and Leon has $3.75. Do they have enough money to buy the pizza?

    Ⓐ Who has the most money?

    Ⓑ Does their money total more or less than $8.95?

    Ⓒ How much does their money plus the cost of the pizza equal?

    Ⓓ What is the average contribution each person is making?

    To solve this problem, you will need to find out how much money they have, then compare that number to the price of the pizza. B is correct. Now try these on your own.

2. Katie played on the computer for 15 minutes. Jahan played 10 minutes longer. How long did she play?

    Ⓐ Which person played 10 minutes longer?

    Ⓑ What was the total amount of time both girls played?

    Ⓒ How long did Jahan play?

    Ⓓ Who played longer, Katie or Jahan?

3. Debbie planted a garden with 42 plants in 6 rows. How many plants were in each row?

   Ⓐ How many plants were in 1 row?

   Ⓑ How many plants were in the garden?

   Ⓒ How much is 42 + 6?

   Ⓓ How many rows were in the garden?

B Write a phrase or sentence to restate the question in your own words.

1. If a faucet drips twice a minute, how many times will it drip in an hour?

2. Heather's class raised $239 with a bake sale. Their goal is $330. How much more do they need to raise?

C Solve. Show your work.

1. Rachel started her homework at 4:05 P.M. and ended at 4:45 P.M. How long did she work on it?

2. Half of the families in David's building have cats. If there are 230 families living there, how many have cats?

3. On Tuesday the temperature was 78° F. The next day it was 55°. How much did the temperature fall?

Understanding the Question 7

LESSON 2

USING CLUES

1. Gather Information
2. Plan
3. Calculate
4. Check

When you solve a story problem, you are like a detective solving a mystery. Like a good detective, you use clues to discover the unknown.

Laura baked 45 cookies and told her brother he could take some to school for his class. When she looked on the plate the next day, there were 28 cookies. How many did her brother take to school?

What we know:	What we need to know:
Laura started with 45 cookies.	How many cookies did her brother take from the 45 she started with?
She ended up with 28 cookies.	

A   Which tells what we need to find out?

1. Raushan and Alex collected 79 cans for a food drive from three classes. Mr. Owen's class donated 16 cans. Ms. Capaldi's class donated 23. Raushan and Alex's class donated the rest. How many cans did Raushan and Alex's class donate?

   Ⓐ the name of the class that collected the most
   Ⓑ the total number of cans collected
   Ⓒ the amount from Raushan and Alex's class
   Ⓓ the amount from Ms. Capaldi's class

   To solve this problem, you will need to find out exactly how many cans that Raushan and Alex's class donated. C is correct. Now try these on your own.

2. Len built a block tower in even columns. He used 49 blocks and made seven rows. How many were in each row?

   Ⓐ the number of blocks in a row
   Ⓑ the number of blocks in a column
   Ⓒ the number of blocks in the tower
   Ⓓ the number of blocks in seven rows

8   Unit 1: Getting Started

3. Ms. Joyce's class left on a field trip at 9:00 A.M. They got back at 11:45 A.M. How long was their trip?

   Ⓐ What time did they return?

   Ⓑ When did they leave on their trip?

   Ⓒ How long is it between 9:00 and 11:45?

   Ⓓ How long is it between 11:45 and 9:00?

B **Write a sentence or phrase to restate the question in your own words.**

1. Camille's family went out of town to see her cousins. They left 12 cans of dog food for her neighbor to feed their dog. When they got back there were 3 cans left. If her dog eats $1\frac{1}{2}$ cans per day, how long did they stay away?

2. Jesse's softball team won 50 percent of its games. How many games did it win if it lost 8 games?

C **Solve. Show your work.**

1. Alma was building a fence for her rabbits. One side of the fence was 10 feet, another was 8 feet. If her fence was shaped like a rectangle, how much more would she need to go all the way around?

2. How many would each person get if you divided 65 chocolate bars among 5 people?

3. Louie practiced the drums for 45 minutes a day. How many minutes is that if he practices for 23 days?

LESSON 3

RECOGNIZING IMPORTANT INFORMATION

1. Gather Information
2. Plan
3. Calculate
4. Check

When a word problem gives a lot of information, look for what is important.

David collects baseball cards. He has 25 cards in his collection. The store where David buys his cards sells 5 cards for $2.00. He wants to get 12 cards for his friend Kiana for his birthday. How much would twelve cards cost?

What information helps solve this problem? What information is not needed for solving the problem?

A Which information is NOT important?

1. Maria was painting a wall. She used purple paint. Each gallon could cover 80 square feet. If the wall was 25 feet long and 12 feet high, how many cans would she need? About how much paint would be left over?

    Ⓐ the length of the wall

    Ⓑ the area a can could cover

    Ⓒ the color of the paint

    Ⓓ the height of the wall

    Most of this information is important to figuring out how much paint is used. However, the color of the paint really doesn't matter. C is NOT important. Now try these on your own.

2. Brandon set up the chairs for Math Night at his school. His teacher expected about 20 people to come so he put out 20 chairs. The program began at 6:45. It turned out that 29 people needed chairs. How many more did Brandon have to set up.

    Ⓐ The program began at 6:45.

    Ⓑ Twenty-nine people needed chairs.

    Ⓒ Brandon put out 20 chairs.

    Ⓓ His teacher expected 20 people.

3. Michelle had 6 friends at her house. Of the group, 4 were girls and 3 were boys. Each could each eat 2 or 3 slices of pizza. No one wanted fewer than 2 slices. Her parents wanted to make sure Michelle's friends would have enough. Each pizza had 8 slices. Would 2 pizzas be enough for the friends and Michelle to have as much as they might want?

   Ⓐ Four were girls and three were boys.   Ⓒ The pizzas had 8 slices.

   Ⓑ Each could eat 2 or 3 slices.   Ⓓ No one wanted fewer than 2 slices.

B List the important information that you need to solve the problem.

1. Danielle bought $4.35 worth of items at the store with a $10 bill. How much change did she get back?

2. The fifth grade sold pies to raise money to help people overseas who had been in an earthquake. At the bake sale, a total of 52 pies were sold. The pies sold for $4.00 each. How much money did the fifth grade raise?

C Solve. Show your work.

1. When resting, Linda's heart beat 32 times in half a minute. At this rate, how many times would it beat in a minute?

2. In the "Go For the Gold" Reading Olympics at Sabrina's school, students could get a gold medal if they read 1000 pages. If she read a new 48-page book, every day for 20 days, how much more would she have to read in the final ten days to make her goal?

3. Tomatoes sold for 49 cents a pound at the bodega near Manny's house. His grandmother gave him $7.00 to shop for the family. How much change would he bring back if he bought 4 pounds of tomatoes and a $1.49 carton of milk?

LESSON 4

CHOOSING AN OPERATION

1 Gather Information
2 Plan
3 Calculate
4 Check

When deciding on a way to solve a story problem, look for key words. The table below shows some common words and phrases that appear in word problems that can be used as clues to help decide which operations to use.

If you see the word(s)...	It usually means...
"in all"	add
"and"	add
"difference"	subtract
"less"	subtract
"share equally"	divide
"equal rows"	multiply
"how much greater"	subtract
"altogether"	add or multiply

A Which best tells what the underlined word or phrase suggests about how to solve the problem?

1. Mario collects butterflies. He found 35 of them last year. He got 48 more so far this year. How many does he have <u>altogether</u>?

   Ⓐ add together both numbers
   Ⓑ divide each by two
   Ⓒ subtract this year's from last year's
   Ⓓ average the two years

   By adding the 35 from last year to the 48 this year, we can calculate how many butterflies Mario has <u>altogether</u>. The correct answer is A. Now try these yourself.

2. <u>How much greater</u>, is the cost of 5 school lunches at $2.75 apiece than the cost of a $12.00 baseball ticket?

   Ⓐ add the price of a lunches to the price of the ticket
   Ⓑ divide the price of the ticket by 5
   Ⓒ subtract the price of the ticket from the price of the lunches
   Ⓓ multiply the cost of the lunches by 5

12   Unit 1: Getting Started

3. How could you <u>share</u> 144 cherries with 12 people so each would get an equal amount?

   Ⓐ add the number of cherries and people

   Ⓑ subtract the people from the cherries

   Ⓒ multiply the people by the cherries

   Ⓓ divide the number of cherries by people

B **Explain what each underlined phrase suggests about how to solve the problem.**

1. Janell had a box of 148 crayons, but she gave 6 to her kindergarten buddy. <u>How much less</u> did she have then?

   "How much less" means:

2. Mr. McLoughlan's class planted flowers in the school garden. If they made <u>5 equal rows of 7</u>, how many flowers did they plant?

   "5 equal rows of 7" means:

C **Solve. Show your work.**

1. <u>How much greater</u> is a 120 square foot room than an 80 square foot room?

2. Lenore had 182 paper clips. Sheng had 377. How much did they have <u>in all</u>?

3. Alma brought pumpkin seeds for her classmates to plant. She had 480. How could she <u>share these equally</u> if there were 26 students in her class? Would she have any left over?

Choosing an Operation   13

LESSON 5

PROBLEMS WITH MORE THAN ONE ANSWER

1. Gather Information
2. Plan
3. Calculate
4. Check

Some problems have more than one answer. From doing problems like these we learn how to organize our thinking and how to use our minds. One way to solve problems like these is to make an organized list.

Which combination of coins adds up to 15 cents?

Ways to make 15 cents
1 dime and 1 nickel
1 dime and 5 pennies
3 nickels
2 nickels and 5 pennies
10 pennies and 1 nickel
15 pennies

Making an organized list might not help you be sure if you have all the possible solutions. For some problems, this might be almost impossible. There are so many ways to make $1.00, for example, that its hard to keep track of them all.

Here are some other helpful strategies to keep in mind:

- Use pictures or objects to help you see solutions.
- Make guesses and then check them afterwards.
- Look for patterns.

A Choose a strategy you might use to solve the problem.

1. Ice cream choices at the snack bar were chocolate, vanilla, and strawberry. Sonia and her sister each got a double cone with two different kinds. What are the possible combinations that they could have chosen?

    Ⓐ use a calculator
    Ⓑ add the two people and three flavors
    Ⓒ use three color tiles and group them
    Ⓓ add up the three kinds of flavors

Using color tiles would help you see how the three different flavors could be grouped. The correct answer is C. Now try these on your own.

2. Alejandro's brother has earned 20 cents by helping. What are the different ways Alejandro can combine coins to pay his brother this money?

    Ⓐ add the coins together
    Ⓑ use a ruler and measure the coins
    Ⓒ use a calculator
    Ⓓ use real coins and record different ways

14   Unit 1: Getting Started

3. Mr. Josloff is making a class garden. He got 200 pounds of peat moss. The peat moss came in 50-pound and 100-pound bags. How many different combinations of bags might he have gotten?

Ⓐ weigh the peat moss
Ⓑ list different combinations that total 200
Ⓒ add 200, 100 and 50 together
Ⓓ use a calculator

B Write a phrase or sentence describing your own strategy for solving the problem.

1. Alma scored 11 points in the basketball game. She scored 3-point shots and 2-point shots. How many of each did she score? Find as many answers as you can.

2. There are 12 wheels at the bike rack. Some are bicycles, some are tricycles. How many of each could there be?

C Solve. Show your work.

1. Chickens walk on 2 feet. Dogs walk on 4 feet. If there are 10 feet on the barnyard floor how many dogs and chickens might be there?

2. At the toy store, there are 3 kinds of toys on a shelf. If there are 3 boxes, how many of each toy could there be?

3. If 3 children are sitting on a bench, what are all the combinations of girls and boys that can sit on that bench?

Problems with More than One Answer

LESSON 6

WRITING AN EQUATION

1 Gather Information
2 Plan
3 Calculate
4 Check

The final step of planning the solution to a story problem is to write an equation.

All equations have:
- Numerals
- Operation signs (+, -, x, ÷)
- and signs such as >, <, = and ≠

Some have symbols that tell you to look for missing numbers such as □, __, ?, or *x*.

They can be written horizontally:

23 + 91 = □

or vertically:

```
   23
 + 91
  ???
```

For example . . .

Add 21 days in April, 17 days in May, and 29 days in June.

21 + 17 + 29 = □

A Choose the equation that matches the sentence.

1. Multiply 17 rows by the 9 seats in each row.
   - Ⓐ 17 + 17 + 9 + 9 = □
   - Ⓑ 9 x □ = 17
   - Ⓒ 17 x □ = 9
   - Ⓓ 17 x 9 = □

   The sentence asks us to multiply 17 x 9 to find the unknown. The correct answer is D. Now try these yourself.

2. Subtract 451 Cowboy's fans from the 1000 in the bleachers.
   - Ⓐ 451 - □ = 1000
   - Ⓑ □ - 451 = 1000
   - Ⓒ 451 - 1000 = □
   - Ⓓ 1000 - 451 = □

16 Unit 1: Getting Started

3. Divide the 333 birds into 3 groups.
   - Ⓐ 333 ÷ 3 = ☐
   - Ⓑ ☐ ÷ 3 = 333
   - Ⓒ 3 ÷ 333 = ☐
   - Ⓓ 3 ÷ ☐ = 333

B Write an equation for the sentence.

1. Multiply the 3 sandwiches by their price, which is $1.75 each.

2. Take half of 1262.

C Solve. Show your work.

1. Divide 360 into 4 equal parts.

2. Add $5 and $9 and subtract it from $37.

3. Subtract 20 from 120, then divide the answer by 4.

UNIT 2   LESSON 1

REVIEW

1. Gather Information
2. Plan
3. Calculate
4. Check

When solving story problems, we start by gathering information. Then we plan a solution.

Ellie and Harriet were organizing the materials for a class picnic. They had 19 plates but were expecting 34 people. How many more plates would they need to get?

Gathering Information:
- What is the important information? they are expecting 34 people; they have 19 plates
- What is missing? the number of plates they still need

Planning a Solution:
- About how many plates will we need? Since 19 is close to 20 it looks like they'll need about ten more, that would give them about 30.
- What would be a method to get the answer? Take the total number of plates they'll need (34) and subtract how many they have already (19).
- Write an equation or number sentence to find the missing information: 34 - 19 = ?

A   Which information is the most important?

1. The sixth grade science fair ran from 9:30 to 11:00. How long did it run?
   - Ⓐ it was a 6th grade fair
   - Ⓑ the fair ended at 11:00
   - Ⓒ the fair started at 9:30
   - Ⓓ both B and C

   The most important information is the starting time and the ending time. So D is correct because it includes both pieces of information. Now try these on your own.

2. John and Alexa were making a block building of 1 inch thick blocks that was 8 blocks wide and 6 blocks high. How many blocks did they use?
   - Ⓐ the blocks were 1 inch thick
   - Ⓑ both C and D
   - Ⓒ the building was 6 blocks high
   - Ⓓ the building was 8 blocks wide

18   Unit 2: Finding Solutions

3. The normal temperature for April 22nd is 56 degrees. The actual temperature is 47. How many degrees is it above or below the normal temperature?

   Ⓐ the date is April 22nd

   Ⓑ both C and D

   Ⓒ water freezes at 32 degrees

   Ⓓ actual temperature is 47; the normal temperature is 56

B Write an equation that will help you find the missing information.

1. Harry is a ferrier who puts shoes on horses. If he puts 4 shoes on each horse and he has 144 shoes, how many horses can he shoe?

2. The fire department uses 600 feet of $2\frac{1}{2}$ inch hose on its truck. Each section of hose is 50 feet. How many sections are used on the truck?

C Solve. Show your work.

1. Lauren wants to give every child at her party 5 prizes in their goody bag. If there are 9 children at the party, how many prizes will she need?

2. There are 320 pistachios in a bag. Of these pistachios, 23 have closed shells or are no good. How many are good to eat?

3. For his family's bike trip, Jeremy is packing water bottles. Each person needs 2 quarts. There are 7 people so how many quarts should he bring?

LESSON 2

CHOOSING A METHOD

1. Gather Information
2. Plan
3. Calculate
4. Check

We can use different methods to find solutions. Ann used four different methods last night when she was doing her homework. She explains why:

Math Problem	What she used	Why she used it
600 x 3 = ?	Mental math	"I knew that 6 x 3 is 18. So 600 x 3 is the same thing, just bigger by two zeros at the end."
42 -29 = ?	Paper and pencil	"It's sometimes quicker to do it on paper than to push the buttons on a calculator."
10 ÷ 2 = ?	Make a model	"I can understand division better sometimes if I can see it, so I took ten counters and divided them into two equal piles."
784 x 892 = ?	Electronic Calculator	"I could do this on paper, but when numbers are this big and they aren't even numbers like 200, I like to use a calculator."

Using different methods can save you time, or help you check your work. While you are learning, your teacher may want you to practice with paper and pencil so that you can develop this skill.

A Which method do you think Ann would use to solve these problems?

1. 10 x 5 = ?

    Ⓐ electronic calculator   Ⓒ make a model

    Ⓑ pencil and paper   Ⓓ mental math

This is a problem that can quickly and easily be solved by mental math. The correct answer is D. Now try these problems on your own.

2. 438 x 647 = ?

    Ⓐ electronic calculator   Ⓒ make a model

    Ⓑ pencil and paper   Ⓓ mental math

20   Unit 2: Finding Solutions

3.   137
    - 89

Ⓐ electronic calculator   Ⓒ make a model

Ⓑ pencil and paper   Ⓓ mental math

B **Choose a method for solving the problem, then explain why you chose it.**

1. The movie theater holds 571 people. If the theater is full for 4 shows, how many people would see it?

2. If the theater is full for 10 shows, how many people would see it?

C **Solve. Show you work.**

1. Albert Einstein was born in 1879 and he died in 1955. How long was his life?

2. If a school lunch costs $2, how much do 25 school lunches cost?

3. How could you share 12 apples in a fair way with 4 people?

Choosing a Method

LESSON 3

MAKING A TABLE

1. Gather Information
2. Plan
3. Calculate
4. Check

Some word problems are easier to solve if you organize your data into a table.

On his birthday, Harry always got quarters from his grandmother. She gave him one quarter for every year old he was. When he was 6 he got $1.50 worth of quarters. Now that he is 9, what will be the total amount of his grandmother's gift?

From the way this problem is set up, we can tell that the amount of money Harry gets depends on the number of years old he is.

Here's what we know so far:

Age/Number of quarters	6	7	8	9
total	$1.50	?	?	?

Since the problem tells us that the amount of money goes up by 25 cents each year, we can complete the table by adding 25 cents every year.

Age/Number of quarters	6	7	8	9
total	$1.50	$1.75	$2.00	$2.25

Harry's grandmother gave him $2.25 on his 9th birthday.

A Use the table below to answer the question.

1. Sharon likes to collect trading cards. Their cost is shown in the table below. How much do 3 cards cost?

cards	1	2	3	4
price	8 cents	16 cents	?	?

   Ⓐ 32 cents   Ⓑ 30 cents   Ⓒ 28 cents   Ⓓ 24 cents

According to the table, one card costs 8 cents, so 3 cards would cost 8 cents + 8 cents + 8 cents. An equation you could write to solve the problem is: 8 cents x 3 = 24 cents. So, D is the correct answer. Now try these on your own.

2. How much do 4 cards cost?

   Ⓐ 32 cents   Ⓑ 40 cents   Ⓒ 28 cents   Ⓓ 38 cents

22   Unit 2: Finding Solutions

3. How much would 5 cards cost?

   Ⓐ 42 cents   Ⓑ 40 cents   Ⓒ 44 cents   Ⓓ 45 cents

B Complete the table.

1.

rides	1	2	3	4
tickets	4 tickets	8 tickets		

2.

cups	2 cups	4 cups	6 cups	8 cups
pints	1 pint	2 pints		

C Solve. Make a table and show all your work.

1. In the school store, erasers cost 15 cents each. How much are 4 erasers? How much are 3?

2. The temperature on May 9th was 52 degrees. It went up 2 degrees per day for the next 5 days. What was the temperature each day?

3. When Marco sets the table he puts out 18 pieces of silverware for the six people in his family. How much should he put out when he is setting the table for his family and one guest? His family and two guests? His family and three guests?

Making a Table

LESSON 4

DOES YOUR ANSWER MAKE SENSE TO YOU?

1. Gather Information
2. Plan
3. Calculate
4. Check

You can estimate to make sure that your answers are in the right neighborhood.

Estimating Sums and Differences

When adding or subtracting 2 or more numbers, round both numbers to the highest place of the smaller number.

Example: when adding

```
  431 — round this number to —   430
        the nearest ten
+  36 — round this to the     — + 40
        nearest ten
        Here's your estimate:    470
```

Estimating Products

When multiplying, round numbers to the nearest ten or hundred. That way you can estimate whether your calculation is close to what it should be.

Example: when multiplying

```
  198 — round this number to —   200
        the nearest hundred
x   4                          x   4
        Mental math can help
        you make this estimate:  800
```

A Choose the estimate you think is most accurate.

1.   469
   + 944

   Ⓐ 1,500   Ⓒ 1,400
   Ⓑ 1,300   Ⓓ 1,200

Rounding 469 to the nearest hundred gives us 500. Rounding 944 to the nearest hundred gives us 900. Adding 900 and 500 gives us 1,400, so C is the correct answer. Now try these by yourself.

2.   149
   - 34

   Ⓐ 100   Ⓒ 130
   Ⓑ 110   Ⓓ 120

24   Unit 2: Finding Solutions

3. 296 × 6 = ?
 Ⓐ 1800
 Ⓑ 2400
 Ⓒ 1900
 Ⓓ 1200

B Make estimates for the calculations shown below. Round the numbers to the nearest ten.

1. 168 + 44 = ?

2. 58 × 9 = ?

C Estimate using rounded numbers. Show your work.

1. Billy's mom sold t-shirts for a school fundraiser. She sold 8 shirts at $9.95. About how much money did she collect?

2. The Tigers scored 41 points and the Rockets scored 28. Roughly how many points were scored in the game?

3. Lukeesha had $1.62 in her piggy bank. She bought something that cost 57 cents. About how much money did she have left?

Does Your Answer Make Sense to You?

1. Gather Information
2. Plan
3. Calculate
4. Check

LESSON 5

USING INVERSE OPERATIONS TO CHECK YOUR ANSWERS

After you have completed your calculations you may want to check by doing them again in a different way.

If you added, you can check by making a subtraction problem with the same numbers.

On Tuesday, 35 students rode on Jacob's bus. 48 rode on Ali's bus. How many students rode on the two buses?

Add to solve:

^{1}
  35 students on Jacob's bus
+ 48 students on Ali's bus
  83 students total

And to check, we can switch the numbers around and make it a subtraction problem:

Start with the total.
Subtract the number of students on one of the buses.
This gives you the number of students on the other bus.

7^1
$\cancel{83}$
- 48
  35

We can also use addition to check a subtraction problem.

David was trying to memorize the 50 state capitals. He already knows 29 of them. How many more does he have to learn?

Subtract to solve: Add:

  50 states to learn   21
- 29 David knows already   + 29
  21 capitals to learn   ? ?

A Which equation could you use to check these solutions?

1. Yvette is 52 inches tall. JoAnne is 46 inches tall. How much taller is Yvette?
   Solution: 52 inches - 46 inches = 6 inches

   Ⓐ 46 - 6 = 40   Ⓑ 52 + 6 = 58   Ⓒ 6 + 46 = 52   Ⓓ 52 + 46 = 98

   In the calculation, 46 is taken away from 52 leaving 6. Answer C is the only choice that has the two smaller numbers added together to make the larger. C is the best choice. Now do the rest on your own.

2. Charles had 372 seeds and planted 96 of them. How many did he have left?
   Solution: 372 seeds total - 96 planted seeds = 278 seeds left

   Ⓐ 278 + 96 = 372   Ⓑ 278 - 96 = 182   Ⓒ 372 + 96 = 468   Ⓓ 372 + 278 = 650

Unit 2: Finding Solutions

3. Laura's 21 inch-long hair had grown down to her waist. She got 9 inches cut off. How long was her hair after her haircut?

   Solution: 21 inches - 9 inches = 12 inches

   Ⓐ 12 - 9 = 3   Ⓑ 9 + 12 = 21   Ⓒ 12 + 9 = 21   Ⓓ both B and C

B Use inverse operations to check the solution. Correct if necessary.

1. Javier has $4.25 but he needs $5.00 to go to the movies. How much more money does he need?

   Solution: $5.00 - $4.25 = $ 0.75

2. Edwardo has to read 78 pages this week for his reading log. He's already read 51. How many more does he have to read?

   Solution: 78 - 51 = 27 pages

3. Aaron's soccer team won 18 games and lost 6. How many games did they play?

   Solution: 18 + 6 = 22 games

4. Irene's had 23 cookies in a bag to bring to school to share on her birthday. Her mom put some more in the bag and she had 49. How many cookies did her mom put in the bag?

   Solution: 49 - 23 = 26 cookies

5. Jerry could hold his breath under water for 43 seconds. His younger brother Artie could hold his breath for 58 seconds. How much longer can Artie hold his breath?

   Solution: 58 - 43 = 5 seconds longer

LESSON 6

SHARING YOUR THINKING

After you've done your calculations and found your answer, you can use words and pictures to show how you got it.

Nino is making a pizza. He cuts it into 8 equal slices. Half of the pizza has broccoli and half is plain cheese. How many slices have broccoli?

Here is how Judi explained how she solved this word problem.

> I knew that:
> - there are 8 slices
> - half are broccoli
> - half are cheese
>
> So since one pizza is 8/8 and 4/8 = 1/2
> 8/8 - 4/8 = 4/8
> 4/8 of the pizza is 4 slices.
> So 4 slices have broccoli.

A Which would help us explain the solution to the word problem?

1. The Gonzalez Building is 26 stories tall. The first 13 floors are shops and the rest are offices. How many floors are offices?

   Ⓐ a tall rectangle with half of it shaded

   Ⓑ a wide rectangle the left side of it shaded

   Ⓒ a circle divided in half

   Ⓓ a triangle divided in half

   To picture a building, a tall rectangle is the best choice. The problem says that the bottom half is shops and the top is offices. A is the best choice. Now do the next two on your own.

2. Jack had 90 comic books in his collection. Then he sold 26 of them. How many were left?

   Ⓐ 90 squares with 26 circles next to them

   Ⓑ 90 circles with 26 of them x'ed out

   Ⓒ 26 squares x'ed out

   Ⓓ 26 circles and 90 squares

28 Unit 2: Finding Solutions

3. Ada had 65 shells, and found 56 more. How many did she have altogether?

   Ⓐ 65 lines with 56 x'ed out
   Ⓑ 65 shells in 56 piles
   Ⓒ 65 lines with 56 circles added
   Ⓓ 65 rows of 56

B **Make a picture that would help you solve these problems.**

1. Michael was making a model of a ship. The real ship is 100 feet tall. The model is 1/50th of the size. How tall is the model?

2. Steve is recyling newspapers. His box holds 16 pounds of paper. If he had 3 boxes, how many pounds of paper could he collect?

C **Solve. Make a diagram and show your work.**

1. Danny is helping to order food for a neighborhood party. He knows that one cake can serve 44 people. If he has another half of a cake, how many people can be served?

2. Justyn had four kinds of beans. How many beans would he have in all if he made a pile of 9 beans for each kind?

3. Rose was making lemonade. She knew that two pints make a quart. If she wanted to make three quarts, how many pints would that be?

Sharing Your Thinking

UNIT 3 LESSON 1

SKILL TUNE-UP: MULTIPLYING AND DIVIDING WHOLE NUMBERS

1. Gather Information
2. Plan
3. Calculate
4. Check

Multiplication and division are related. Think about this fact family.

Multiplication	Division
3 x 7 = 21	21 ÷ 3 = 7
7 x 3 = 21	21 ÷ 7 = 3

A List the answers.

1. What facts based on 2, 9, and 18 are in the same fact family?

Multiplication	Division

Number sentences in a fact family use the same numbers.

2 x 9 = 18; 9 x 2 = 18; 18 ÷ 2 = 9; 18 ÷ 9 = 2

Try the next on your own.

2. What facts based on 6, 5 and 30 are in the same fact family?

Multiplication	Division

3. What facts based on 2, 6, and 12 are in the same fact family?

Multiplication	Division

4. What facts based on 5, 11, and 55 are in the same fact family?

Multiplication	Division

Unit 3: Multiplying and Dividing Whole Numbers

5. What facts based on 4, 5, and 20 are in the same family?

Multiplication	Division

6. What two multiplication number sentences use 12, 6, and 72?

7. What two division number sentences use 72, 12, and 6?

B For each set of numbers, list the multiplication and division number sentences that make a fact family.

1. 7, 6, 42

2. 9, 7, 63

3. 7, 5, 35

4. 4, 8, 32

5. 6, 8, 48

6. 8, 9, 72

7. 7, 9, 63

8. 4, 9, 36

9. 6, 4, 24

10. 7, 8, 56

11. 5, 12, 60

12. 8, 12, 96

Skill Tune-up: Multiplying and Dividing Whole Numbers 31

LESSON 2

STRATEGY: DRAW A PICTURE

1. Gather Information
2. Plan
3. Calculate
4. Check

When you want to solve a word problem involving multiplication or division, drawing a picture can help.

Desiree was planning an art show of her drawings. She had 32 drawings that she put in four rows. How many drawings were in each row?

Which could best help you solve the problem?

A

X X X X X X X X X X X X X X X X
X X X X X X X X X X X X X X X X

B

X X X X X X X X
X X X X X X X X
X X X X X X X X
X X X X X X X X

Picture B is best because the X's are organized into 4 rows, and it's easy to see that there are 8 in each row.

A Which drawing could help you solve the problem?

1. There are five people in LeRoy's family. How could they divide 25 muffins so that everyone gets an equal amount?

    Ⓐ five rows of 5 muffins
    Ⓑ 25 muffins
    Ⓒ 25 muffins added to 5 people
    Ⓓ 25 muffins organized into five groups

The solution to this word problem is 25 ÷ 5 = 5. A drawing showing 25 muffins in five piles would help you solve this word problem. The correct answer is D. Now try these problems on your own.

2. Terry was helping her dad wash the dishes. 6 cups could fit on every shelf in the kitchen. If there were 4 shelves, how many cups could they put on the shelves?

    Ⓐ 4 shelves with 6 cups on each
    Ⓑ 6 shelves with 4 cups
    Ⓒ 4 shelves and 6 cups
    Ⓓ 6 shelves with 4 cups on each

Unit 3: Multiplying and Dividing Whole Numbers

3. There are 12 inches in a foot. How many inches long are 4 rulers that are each a foot long?

   Ⓐ four rulers totaling 12 inches

   Ⓑ 12 rulers equalling 4 feet

   Ⓒ 4 rulers that are each 12 inches

   Ⓓ 4 rulers totalling 12 feet

B Make a drawing and solve the problem. Show your work.

1. Katelyn was helping her mom put down tile in the hallway of their apartment. One section of the floor was 5 tiles long and 4 tiles wide. How many floor tiles did they use?

2. Patrick takes one vitamin tablet a day. His vitamins come in bottles with 30 vitamin tablets. How many vitamins are in three bottles?

3. What is a fair way to divide 23 cherries three ways? How many would be left over?

4. Lloyd needed 48 colored pencils. There were 12 in each package. How many packages did he have to buy to get enough pencils?

Strategy: Draw a Picture   33

LESSON 3

STRATEGY: MAKE A LIST

1 Gather Information
2 Plan
3 Calculate
4 Check

Some word problems are easier to solve if you organize your data into a list. Here's an example.

The fourth grade is taking a bus to a field trip. Three children can fit in each seat on the bus. There are 16 seats on a bus. 58 children are in the fourth grade. Can the entire fourth grade fit on one bus?

What We Know	What Need to Know/How to Find it Out
• There are 58 children in the 4th grade	• How many children could fit on the bus
• The bus has 16 seats	• Multiply the number of seats (16) by the number
• Three children sit in each seat	of children who can sit in each seat (3)

A Choose the best answer to the questions about the problem.

A bag has 320 almonds in it. Kyla's class has 24 students. Is there enough in the bag to give everyone in Kyla's class exactly a dozen almonds?

1. What information does the problem give us?

   Ⓐ how many almonds will be eaten

   Ⓑ how many students are in the class

   Ⓒ how many students are in the school

   Ⓓ how many bags of almonds there are in all

The problem does not tell how many almonds will be eaten or how many students are in the whole school. It does not say how many bags of almonds there are in all. It does say that there are 24 students in Kyla's class, so, B is the correct answer. Now try these on your own.

2. What do we need to find out?

   Ⓐ Are enough almonds to give 12 to each student?

   Ⓑ Does the bag have exactly 320 almonds?

   Ⓒ How many students would it take to use all 320 almonds?

   Ⓓ How many bags would it take to hold all 320 almonds?

Unit 3: Multiplying and Dividing Whole Numbers

3. How can we find a solution to this problem?

   Ⓐ subtract 12 from 24 then compare that total to 320

   Ⓑ divide 320 by 24 and compare that total to 0

   Ⓒ multiply 12 x 24 and compare that total to 320

   Ⓓ add 12 and 24 and multiply that total by 12

B Complete the tables.

1. Ms. Valerio is going to teach a group of 8 third graders how to knit. Each child needs a ball of yarn at least 20 yards long. How many yards of yarn will she need?

What We Know	What Need to Know/How to Find it Out
•	•
•	•

2. To play a card game, Jakki deals down 42 cards into 6 equal rows. How many cards does she use in each row?

What We Know	What Need to Know/How to Find it Out
•	•
•	•

C Solve. Make a table of known and unknown information and solve these problems. Show all your work.

1. Ms. Saltz's class hope to raise $30 with a bake sale. If baked goods sell for 25 cents apiece, how many items will they have to sell?

2. Renata's class was going to have a pizza party. They got 7 pizzas. Each pizza has 8 slices. There were 23 children and 2 adults. How many pieces could each person have?

Strategy: Make a List   35

LESSON 4

STRATEGY: WORK BACKWARDS

1. Gather Information
2. Plan
3. Calculate
4. Check

For some word problems, you may have to figure out an answer by tracing steps backward. Here's an example:

Eddie spent $8 on school lunches this week. He brought lunch from home on Monday. On the other days, he spent the same amount. How much did he spend each day?

Before we can calculate this, we need to find the total number of days he bought school lunch:

~~Monday~~ Tuesday Wednesday
Thursday Friday = 4 days

Then we can take the total amount of money he spent during the week.

($ 8)

We can use the information we gathered to divide the total amount of money he spent by the number of days he bought lunch.

$8 ÷ 4 days = $2.00 each day

A Choose the letter of the correct answer.

1. Kyle's friend gave him a gift certificate for a book store. He spent $4 on a science book. Then he got a baseball book for $5 and a coin book for $6. He had $10 credit left on his gift certificate. What was the original value of the gift certificate?

   Ⓐ add $4 + $5 + $6 + $10
   Ⓑ add $5 + $6 + $10
   Ⓒ subtract $10 from $4 + $5 + $6
   Ⓓ subtract $6 - $5 - $4 from $10

   We know that Kyle's purchases left him with ten dollars credit. If we add up the total of the purchases and the amount of the credit, this will give the original value of the gift certificate. The correct answer is A. Now try these problems on your own.

2. Every week, JoAnne and her brother and sister each put 25 cents in a piggy bank. They now have $3.25 in the bank. How much more do they need to have to earn $10?

   Ⓐ add: $10.00 + $3.25
   Ⓑ add: $0.75 + $3.25
   Ⓒ subtract: $10.00 - $0.75
   Ⓓ subtract: $10.00 - $3.25

Unit 3: Multiplying and Dividing Whole Numbers

3. Lauren's mom gave her $20 and told her to get some cans of cat food for the family cat, Statue. Lauren saw a special at the store of 3 cans for a dollar. She brought back $5 in change. How much did she spend on cat food?

   Ⓐ subtract: $20 - $5
   Ⓑ subtract: $20 - 3
   Ⓒ add: $5 + 3
   Ⓓ divide: $20 ÷ 3

B. Write a few words telling about a first step for solving the problem.

1. At the end of the fix-up day, there were 9 cans of paint left over. The helpers used 23 cans of paint in the morning and twice as many in the afternoon. How many did they start with?

2. Katie and her friends took in $22.25 at their tag sale table. After paying a $5.00 table fee, they split the money five ways. How much did each person receive?

C. Solve. Show your work on another piece of paper.

1. Sarah's hair is twice as long as Lizzie's. Lizzie is 5 inches longer than Zach. Zach's hair is three inches long. How long is Sarah's hair?

2. Norah's softball team scored 6 runs in their first game. In their second game they scored twice as many runs. Then, they made the same number of runs in the third game. How many runs did they score in their three games?

3. Jawan sold raffle tickets to raise money for his school orchestra. The tickets cost $2 each. When Jawan was done selling, he had $18 and 6 unsold tickets. How many tickets did he begin with?

Strategy: Work Backwards

LESSON 5

CHOOSE A STRATEGY

1. Gather Information
2. Plan
3. Calculate
4. Check

Here are some useful strategies for solving word problems involving multiplying and dividing whole numbers:

Name of Strategy	How it Helps	Where it Helps
Draw a diagram	helps you see the relationships in the problem so you can decide how to find missing information	when you need help visualizing what the problem is asking
Make a list	helps summarize the information that the problem gives you	when you want to organize the information you found so far in a word problem before you begin to solve it
Look for a pattern	helps you find a relationship among a series of numbers, such as 2, 4, 6, . . . (going up by +2)	when you think you can use a pattern to predict other numbers
Work backwards	helps you find a good first step when solving 2 or 3 step problems	when a complicated word problem requires more than one step
Use objects	helps you see a problem situation with objects such as linker cubes or paper clips to stand in for the items in a word problem	when you need to organize the same objects in different ways in order to solve a problem

A Which strategy would be most helpful in solving the problem?

1. Each day, for a week, Sam bought twice as many baseball cards as he did the day before. On the first day he bought 1 card. On the second day, he bought 2 cards. On the third day, he bought 4 cards. How many cards did he have at the end of the seventh day?

   Ⓐ subtract backwards
   Ⓑ make a list
   Ⓒ look for a pattern
   Ⓓ work backwards

   The number of cards increases by a regular pattern, so C is the correct answer. Now try these by yourself.

38   Unit 3: Multiplying and Dividing Whole Numbers

2. How many beans will be left over if we try to divide 260 of them fairly among 8 people?

   Ⓐ look for a pattern
   Ⓑ work backwards
   Ⓒ make a list
   Ⓓ use objects

3. Martha made a pie and cut it into 12 pieces. How many different people can each get two slices?

   Ⓐ work backwards
   Ⓑ write an equation
   Ⓒ look for a pattern
   Ⓓ draw a diagram

B Name the strategy you would use to solve the problem. Explain why you chose it.

1. When there were 120 people expected at the spaghetti dinner, the sixth graders used 20 pounds of spaghetti. When there were 240 people, they used 40 pounds. How many pounds would they need to feed 360 people?

2. Lia planted 5 rows of marigold seeds, with 8 seeds in each row. Thirty-six plants came up. How many didn't survive?

C Use one of the strategies to solve these problems. Show your work on another piece of paper.

1. For the first day of school, Ms. Washington put 3 pencils on every student's desk. She had 28 students. How many pencils did she need?

2. A football team scored 44 points. It scored 30 points by touchdowns and 5 points by kicking points after touchdowns. It scored the rest on field goals 2 worth 3 points each. How many points did it score on field goals?

3. Twenty-three people parked in the parking lot. There was a $4 parking fee. 10 people got to park for free because they were helpers. How much money did the parking lot attendant collect?

Choose a Strategy

LESSON 6

MULTI-STEP STORY PROBLEMS

1. Gather Information
2. Plan
3. Calculate
4. Check

Some problems require more than one step. Here's an example.

Mr. Fernwald's class made bookmarks from handmade paper. They sold 37 to 6th graders and 44 more to 5th graders. If the bookmarks sold for 50 cents each, how much money did Mr. Fernwald's class collect?

Step 1: We combine the number of bookmarks sold.

37 + 44 = 81

Step 2: We multiply that total number of bookmarks by 50 cents.

81 x $0.50 = $40.50

A Which lists the proper steps for solving the word problem?

1. Kiara had 43 nickels. James had 14 quarters. How much money did they have altogether?

   Ⓐ add, then divide the totals

   Ⓑ subtract, then multiply the totals

   Ⓒ add, then multiply the totals

   Ⓓ multiply, then add the totals

   To solve this problem, you would need to multiply 43 x 5 cents and 14 x 25 cents, then add those two totals. D is the best choice. Now do the next two on your own.

2. At multi-ethnic dinner night, families used 116 chopsticks and 86 forks. Since each person got either one fork or two chopsticks, how many people ate dinner?

   Ⓐ add 116 and 86

   Ⓑ add 116 and 86, then multiply by 2

   Ⓒ divide 116 by 2, then add 86

   Ⓓ multiply 86 by 2, then add 116

Unit 3: Multiplying and Dividing Whole Numbers

3. Mr. Green wanted to spend $15 per student on school supplies. His school gave him $300. He had 22 students. How much more money would he need to spend?

   Ⓐ divide 300 by 22, then add $15

   Ⓑ multiply $15 by 22, then subtract $300

   Ⓒ add $15 and $300, then divide by 22

   Ⓓ subtract $15 from 22, then divide 300 by that number

B List the steps you would use to solve the problem.

1. Wood for shelves costs $4 a foot. Wilson needs a 6 foot shelf and a 10 foot shelf. How much money does he need?

2. Ms. Barbezat's class of 28 students was going to the planetarium. The PTO paid $20 toward class field trips. The trip cost 76 dollars. How much did each student have to pay?

C Solve. Show your work on another piece of paper.

1. Jordan's class bought 21 books that cost $2 and 14 books that cost $3 from the book club in January. How much did they spend on their order?

2. Alicia helped collect recyclable bottles. The big bottles were worth 10 cents and the cans were worth 5 cents. She collected 47 bottles and 19 cans. How much money did she collect?

3. Ms. Lambert's class had a science fair. Ms. Raines' class got to visit for 40 minutes. They got to spend 8 minutes at each exhibit, then a bell would ring. How many exhibits did Ms. Raines' class get to see? How much extra time did they have left over at the end?

UNIT 4   LESSON 1

SKILL TUNE-UP: PATTERNS AND RELATIONSHIPS

1. Gather Information
2. Plan
3. Calculate
4. Check

To solve some problems, look for a pattern.

For example, look at these numbers: 1, 2, 2, 3, 3, 3, 4, 4, 4, . . .

- Step 1: Make a guess about a rule. (Oh, I see. Each number repeats as many times as its name.)
- Step 2: Test the rule. (There are 2 twos and 3 threes.)
- Step 3: Use the rule to predict the next numbers in the series. (There are 3 fours so far. The next number should be 4.).

Look for patterns and relationships to help you predict what the next number in a sequence will be.

A   Which correctly continues the sequence of numbers?

1. 2, 4, 8, 16, __, __, __, __
   - Ⓐ 17, 18, 19, 20
   - Ⓑ 18, 19, 20, 22
   - Ⓒ 18, 20, 22, 24
   - Ⓓ 32, 64, 128, 256

The correct answer is D because each number is the double of the one before it. Now try these on your own.

2. 3, 6, 9, 12, . . .
   - Ⓐ 14, 16, 18, 20
   - Ⓑ 15, 18, 21, 24
   - Ⓒ 13, 14, 15, 16
   - Ⓓ 14, 17, 20, 23

Unit 4: Patterns and Relationships

3. 11, 22, 33, 44, …
   - Ⓐ 55, 66, 77, 88
   - Ⓑ 50, 51, 52, 53
   - Ⓒ 54, 64, 74, 84
   - Ⓓ 45, 46, 47, 48

B List the next 3 numbers in the sequence.

1. 0, 50, 100, 150…

2. 1, 1, 3, 3, 5, 5, 7, 7, …

C Write a list of 6 numbers that follow the rule.

1. Start at 0. Count by 25s for the next 5 numbers.

2. Start at 90. Write the next 5 numbers counting backwards by 5s.

3. Start at 2. For the next number go up by 3. For the next number go down by 1. Keep following this pattern until you have 6 numbers.

LESSON 2

STRATEGY: DESCRIBE THE CHANGES

1. Gather Information
2. Plan
3. Calculate
4. Check

Look at this series of critters and think about how they are changing as we move down the line.

Now in a few words, what do you notice?

ant
mouse
squirrel
cat
goat

The size of the words also shows the size of each animal.

A Which best describes how the numbers or letters in the list change?

1. 20, 40, 60, 80 . . .

   Ⓐ the numbers double
   Ⓑ each number goes up by 10
   Ⓒ each number goes up by 20
   Ⓓ they are all even numbers

   The numbers are even, but there are only certain even numbers included in the list. Choice C is more specific and is also true. It is the best choice. Now try these on your own.

2. Z, Z, Y, Y, X, X, W, W . . .

   Ⓐ going up the alphabet
   Ⓑ going down the alphabet
   Ⓒ going up the alphabet in double letters
   Ⓓ going down the alphabet in double letters

44 Unit 4: Patterns and Relationships

3. 5, 7, 9, 11...

    Ⓐ odd numbers greater than 3

    Ⓑ counting by 2s

    Ⓒ counting up by fives

    Ⓓ going up by 3 each time

B Write a few words to describe the changes you see in these patterns.

1.
```
   *
  ***
 *****
*******
*********
```

2. ▲   ■   ⬟   ⬢

C Continue the pattern in the blanks, then describe the pattern you found in words.

1. 7, 14, 21, ___, ___, ...

2. X, X, O, X, X, O, ___, ___, ...

3. 2, 5, 8, ___, ___, ...

Strategy: Describe the Changes

LESSON 3

STRATEGY: MAKE A PREDICTION CHART

1. Gather Information
2. Plan
3. Calculate
4. Check

Here is a problem that Alan solved:

At the library, first graders can take out two books, second graders can take out four, and third graders can take out six. From what you know so far, how many books do you think sixth graders could take out?

First, Alan made a chart to help him see the big picture.

What do you notice about the ways the numbers on the top part of the chart and the numbers on the bottom part of the chart go together?

Here's what Alan wrote in his math journal:

"The bottom numbers are twice as big as the top ones.

Every time the top number goes up one, the bottom number goes up two."

Grade	1	2	3	4	5	6
# of books	2	4	6	...	...	...

A Use the chart to make these predictions.

1. How many books do you think sixth graders can take out?
   - Ⓐ 6
   - Ⓑ 7
   - Ⓒ 9
   - Ⓓ 12

From the chart, we know that kids can take out twice as many books as their grade. Sixth graders can take out 6 x 2 books. 6 x 2 = 12. So D is the right answer. Now answer these questions on your own.

2. How many books do you think fourth graders can take out?
   - Ⓐ 7
   - Ⓑ 8
   - Ⓒ 6
   - Ⓓ 4

Unit 4: Patterns and Relationships

3. If the pattern keeps going to seventh grade, how many books can seventh graders take out?

   Ⓐ 7   Ⓒ 12
   Ⓑ 14   Ⓓ 13

B Complete these prediction charts.

1.

bicycles	5	10	15	20	25	30
bicycle tires needed	10	20	30			

2.

10 packs	$9.95	$19.00	$28.50			
single tapes	$0.99	$1.90	$2.85	$3.80	$4.75	$5.70

C Solve. Make a prediction chart for the problem.

1. Dameon was keeping track of his savings for a basketball. He earns an allowance of $1.50 per week. How long would it take him to buy a $10 basketball?

2. Joe collected signatures to help start a recycling center in his town. He could get about 12 signatures in an hour. How many hours would it take him to get 60 signatures?

3. Jeremy was planning a summer bicycle trip with his family. They would need about three quarts of water for each hour of the trip. If the trip was 6 hours, how much water would they drink?

LESSON 4

STRATEGY: USE AN ANALOGY

1. Gather Information
2. Plan
3. Calculate
4. Check

Here are two problems that Loretta's teacher gave the class to solve.

Problem A	Problem B
A giraffe was 14 feet tall. A horse standing next to the giraffe was 6 feet tall. How much taller was the giraffe?	Micha had 231 stones in his mineral collection. Brian had 144 stones in his collection. How many more stones did Micha have?

Even though the problems are different, they can be solved the same way.

Loretta described how in her math journal:

"Both problems are looking at two amounts of things and asking which is more. The way I would solve them is to take the bigger number and subtract the smaller number from it."

In the first problem she took the bigger number (14 feet) and subtracted the smaller one (6 feet)   14 − 6 = 8

So the giraffe was 8 feet taller than the horse.

In the second problem, she did the same thing.

She took the bigger number (231 stones) and subtracted the smaller one (144 stones)   231 − 144 = 87

Micha had 87 more stones than Brian.

A   How is Problem A like Problem B in the examples below?

1.

Problem A	Problem B
Mr. Sullivan's class walked a half mile to the museum, $1\frac{1}{2}$ miles to the park, then 1 mile back to school. How many miles did they walk altogether?	The Hawks scored 23 points in the first quarter, 14 points in the second quarter, 20 points in the third quarter and 17 points in the fourth quarter. How many points did they score in the game?

Ⓐ both use subtraction to find a total

Ⓑ both involve adding fractions

Ⓒ they both use addition to find a total of several numbers

Ⓓ both are about sports

Both problems can be solved by adding a string of numbers together, so C is the correct answer. Now try these on your own.

48   Unit 4: Patterns and Relationships

2. | Problem A | Problem B |
    |---|---|
    | Each day, Dora got a surprise in her lunchbox from her mom. On the first day, she got two pennies, on the second day, she got two nickels, on the third day she got two dimes. What do you think she got on the fourth day? | Willie's scores on the spelling test kept going up throughout the school year in regular jumps. On his first test he got 14, on the second he got 16, on the third he got 18. By this pattern what do you think he would have gotten on the fifth test? |

Ⓐ each problem is solved by adding the totals

Ⓑ both problems are solved by finding patterns

Ⓒ the two problems can be solved by subtracting

Ⓓ they both have fractions that add up to whole numbers

B   Write down your ideas.

1. In your own words, tell how Problem A is like Problem B.

Problem A	Problem B
Pascual was buying supplies for school. He had $5.00 to spend. He bought a notebook for $1.25, a pencil case for 90 cents, and 3 ball point pens for 40 cents each. How much change did he have left over?	Lori went to the sandwich shop with her mom. She had a vegetarian sub for $2.25, an ice tea for 90 cents and a bag of chips for 75 cents. She gave the counter person a $10.00 bill. How much did she get back?

   _____

   _____

2. In the box next to Problem B, write and solve a similar problem of your own with different numbers and a different story, but the same kind of math thinking.

Problem A	Problem B
Steve lives three miles from school. How many miles does he ride his bike in 5 days if he rides his bike each way? 3 miles x 2 = 6 miles per day 6 miles x 5 days = 30 miles per week	

3. | Problem A | Problem B |
    |---|---|
    | Abraham Lincoln was born in 1809 and died in 1865. How old was he when he died?
1865-1809= 56 years | |

Strategy: Use an Analogy   49

LESSON 5

CHOOSE A STRATEGY

1. Gather Information
2. Plan
3. Calculate
4. Check

Here are some strategies you may find helpful when solving word problems about patterns and relationships.

Name of Strategy	Problem	How to use the strategy
Make a Table	If there are 5 French Francs to a U.S. dollar, how many Francs is $5? $10? $15?	Make a table with Francs on the top and U.S. dollars on the bottom. Multiply the Francs by 5 to get the dollar amounts.
Draw a Picture	A 12" pizza will make 6 slices that are 2" wide. How many slices will an 18" pizza make?	Draw a circle divided into 6. Draw a circle 1/3 bigger and see how many slices of the same size it can make.
Use Objects	For every 3 pencils, Kate had one eraser. If she had 48 pencils, how many erasers did she have?	Place 48 pencils on your desk. Divide them into piles of 3. Put an eraser on each pile. Count the number of erasers you used.
Act it Out	Willy took 20 steps down the hall to the Art room, then double that many to get to the library. How many steps did he take in all?	Walk and count off 20 steps Then double that number (40). Record the number of steps you take.
Do the Problem Backwards	Jeanne had a coupon book for movie tickets. The tickets cost $3. The book had $21 worth of tickets when it was new. Now it was $9. How many tickets did she use?	Start with 21. Subtract $3 multiple times until you get to $9. 21-3-3-3-3=9. She used 4 tickets.

A   Choose the best strategy for solving the problems below.

1. Dina can do 36 jumping jacks in a minute. How many can she do in 2 minutes?

    Ⓐ Act it out by jumping 36 times, then counting 36 more.

    Ⓑ Draw a picture of someone doing jumping jacks.

    Ⓒ Work backwards, counting from 2 minutes to 1 minute.

    Ⓓ Use objects to represent minutes and jacks.

   Acting it out is the best of these methods for this problem. The correct answer is A. Now try these on your own.

Unit 4: Patterns and Relationships

2. On the fourth grade walking trip, every group of students needed to go with one adult. 12 students had 2 adults. If there were 36 students, how many adults attended?

  Ⓐ Act it out by gathering 36 students

  Ⓑ Do the problem backwards by starting with the correct number of adults needed

  Ⓒ Draw a picture of 36 students

  Ⓓ Make a table comparing the number of students and adults

B Explain how you might use the strategy listed to solve each problem.

1. Ann had 48 pieces of paper to make mini-books. She used paper clips to bundle them. Each book needed 3 sheets. How many paper clips did she need? (use objects)

2. A car can go 263 miles on a tank of gas. Its gas tank holds 11 gallons. How many miles can it go on 8 gallons? 6 gallons? (make a table)

C Pick a strategy to solve the problem. Solve and show your work.

1. Every year, a tree in front of Leo's school grew 16 mm. It was 70 mm tall when he began kindergarten. How tall was it when he began sixth grade?

2. Danny takes a vitamin tablet every morning. About how many months would it take him to finish a 90-tablet bottle?

3. Cindy made trains of linker cubes. Each train had 13 cubes. She had 143 cubes. How long were the trains?

1. Gather Information
2. Plan
3. Calculate
4. Check

LESSON 6

MULTI-STEP STORY PROBLEMS

To solve multi-step problems, you will need to do more than one calculation to arrive at your solution. Here's an example:

> Jake's job was to buy bread for the peanut butter sandwiches the fourth grade was going to eat on a picnic. Each loaf had 17 slices. How many loaves of bread would he need to make 100 sandwiches?

The first step: 100 sandwiches x 2 = 200 pieces of bread.

The next step: 200 ÷ 17 = 11 loaves with 13 slices left over.

A Identify the next step in solving the problem.

1. If there are 3 tennis balls in a package, how many more packages will Ms. Taylor, the gym teacher, need if she has 12 packages but needs to have 40 tennis balls.

   The first step: Multiply 12 packages x 3 = 36 balls

   The next step:

   Ⓐ subtract 40 - 12 = 28 balls
   Ⓑ add 12 + 40 = 52 balls
   Ⓒ subtract 40 - 36 = 4 balls
   Ⓓ divide 36 ÷ 3 = 12 packages

   Subtracting the number of tennis balls she has from the number she wants is the next step. C is the best choice. Now try these on your own.

2. Miranda brought up 4 gallons of ice cream from the freezer. Junior added 3 more quarts. How many quarts did they bring up? (Remember, there are 4 quarts in a gallon.)

   The first step: Multiply 4 gallons x 4 = 16

   The next step:

   Ⓐ Add 3 + 16 = 19 quarts
   Ⓑ Add 3 + 4 = 7 gallons
   Ⓒ Multiply 16 x 3 = 48 quarts
   Ⓓ Divide 16 ÷ 3 = 5 1/3 quarts

Unit 4: Patterns and Relationships

3. Sarah helped her dad load cartons of 12 eggs. They loaded 70 cartons. They found out later, that in each carton, 1 egg was no good. How many good eggs did they load?

First step: 70 cartons x 12 = 840 eggs

Next step:

Ⓐ add 70 eggs to the 840 in the cartons

Ⓑ subtract 12 eggs from the 840 in the cartons

Ⓒ subtract 1 egg from the 840 in the cartons

Ⓓ subtract 70 bad eggs from the 840 in the cartons

B List the steps in solving the problem.

1. Bus drivers at Randy's school earn $16,000 per year. Each year, their salary goes up $1,500. How much more will they make 5 years from now than they will make next year?

2. 110 Japanese Yen is worth one U.S. dollar. Jordan started his trip to Japan with $20. He spent 770 Yen on souvenirs. How many Yen did he have left at the end of his trip?

C Solve. Show your work.

1. Doreen was making a pattern with tiles. The pattern was: blue, blue, red, red, green, blue, blue, red, red, green, blue. She ran out of red and had to use purple. What were the colors of the next five tiles?

2. According to the rain gauge, it rained two inches per hour during the storm on May 6th. If the storm went on for 5 hours, how much more rain fell than the 8 inches that fell on April 26th?

3. A recycling bag holds 35 lbs. of paper. Bags come in packages of 10. Is one package of bags enough to recycle 278 lbs. of waste paper?

Multi-step Story Problems   53

UNIT 5 LESSON 1
SKILL TUNE-UP: TIME AND DISTANCE

A Choose the correct answer.

1. The distance from New York City to Washington D.C. is about 240 miles. If a bus is going 60 miles per hour, how long will the trip take?
   - Ⓐ about 3 hours
   - Ⓑ about 4 hours
   - Ⓒ about $4\frac{1}{2}$ hours
   - Ⓓ about 6 hours

   In time and distance problems, we multiply the time by the rate of speed to find the distance traveled.

   If we want to find the time, we can divide the total number of miles traveled by the speed.

   240 ÷ 60 miles in one hour = 4 hours

   The correct answer is B. Now try these on your own.

2. How long would it take to make a 150 mile trip from Richmond, Virginia to Baltimore, Maryland, if the driver keeps a steady speed of 50 miles per hour?
   - Ⓐ 2 hours
   - Ⓑ 3 hours
   - Ⓒ $4\frac{1}{2}$ hours
   - Ⓓ $2\frac{1}{2}$ hours

3. Marina's family drove 1000 miles from Chicago to Jacksonville, Florida. How many hours of driving did it take them if they traveled at 50 miles per hour?
   - Ⓐ 20 hours
   - Ⓑ 15 hours
   - Ⓒ 50 hours
   - Ⓓ 25 hours

B Solve. Show your work.

1. An ultra-marathon runner ran 11 miles per hour for 3 hours. How far did she run?

2. Mychaela went on a bike ride with her family. They traveled at 4 miles per hour for 3 hours. How far did they go?

3. Daryl's trip to school is 2 miles each way. How many miles does he travel to and from school in a week?

4. The distance from Chicago to Milwaukee is 90 miles. If a bus goes 45 miles per hour, how long will it take to get there?

5. A jet plane travels at 575 miles per hour. If the US is 2300 miles wide, how long would it take this plane to travel from one coast to the other?

LESSON 2

STRATEGY: MAKE A TIME AND DISTANCE GRAPH

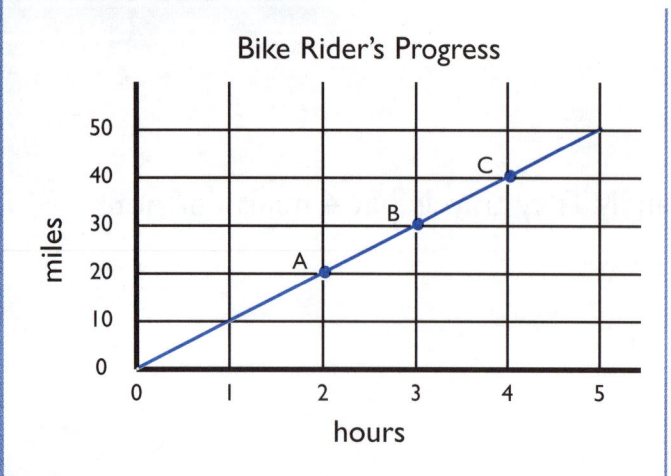

1. Gather Information
2. Plan
3. Calculate
4. Check

Bike Rider's Progress

This is a graph that shows how many miles a bicycle rider traveled during a 5 hour ride. The numbers along the side show the number of miles she traveled. The numbers along the bottom show how much time she has been riding.

If you look at any point along the line, you can see how long and how far the rider had gone. For example, at point A, the rider had cycled for 2 hours, and had covered 20 miles.

A Use the graph to help you choose the correct answer.

1. How many hours did it take the cyclist to go 40 miles?
   - Ⓐ 4 hours
   - Ⓑ 2 hours
   - Ⓒ 3 hours
   - Ⓓ 5 hours

Point C on the line is at 40 miles and 4 hours. A is the correct answer. Now try these on your own.

2. After 2 hours, how many miles had the cyclist gone?
   - Ⓐ 30 miles
   - Ⓑ 20 miles
   - Ⓒ 25 miles
   - Ⓓ 50 miles

3. How many miles did the cyclist go after $2\frac{1}{2}$ hours?
   - Ⓐ 10 miles
   - Ⓑ 1 mile
   - Ⓒ 20 miles
   - Ⓓ 25 miles

Unit 5: Time and Distance

B Use the graph to answer these questions.

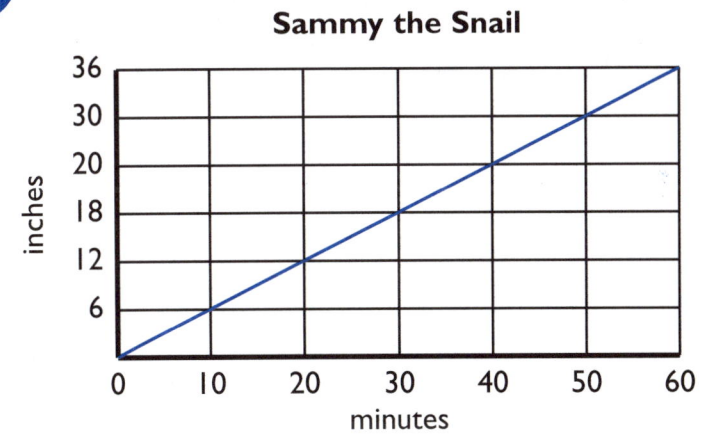

Sammy the Snail

Sammy the Snail takes 10 minutes to move 6 inches. Here's how he moved over an hour.

1. How many minutes did it take Sammy to move 18 inches?

2. After 50 minutes, how far did he go?

C Solve. Show your work.

1. A plane flew at 150 miles per hour. In 1 hour, the plane traveled 150 miles. In 4 hours it flew 600 miles. Make a graph showing how far it flew in 6 hours.

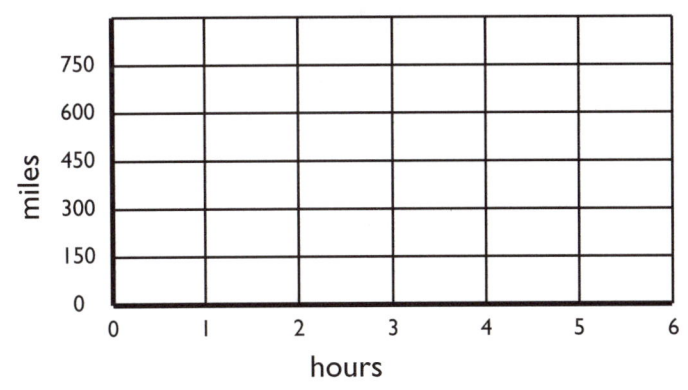

2. How far did it fly after three hours?

3. How long did it take the plane to fly 300 miles?

Strategy: Make a Time and Distance Graph   57

LESSON 3

STRATEGY: DRAW A CLOCK FACE

Starting time   **Finishing time**

Here is how long it took Rajiv and his father to hike 7 miles.

A Choose the correct answer.

1. How long did it take them?

   Ⓐ 1 hour and 20 minutes   Ⓒ 2 hours and 40 minutes

   Ⓑ 2 hours   Ⓓ 2 hours and 20 minutes

The finish time of 2:20 is 2 hours later than the 12:20 starting time. B is the correct answer. Now try these on your own.

2. How long did it take them to hike 10 miles if they finished at 3:10?

   Ⓐ 2 hours and 45 minutes   Ⓒ 2 hours and 50 minutes

   Ⓑ 3 hours and 30 minutes   Ⓓ 3 hours and 50 minutes

3. Which clock face shows the time three hours and 30 minutes after the starting time?

   Ⓐ

   Ⓑ

   Ⓒ

   Ⓓ

58   Unit 5: Time and Distance

B Draw hands on the clock face to show the starting and finishing times.

1.

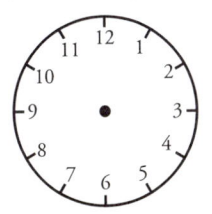

4:17 starting time 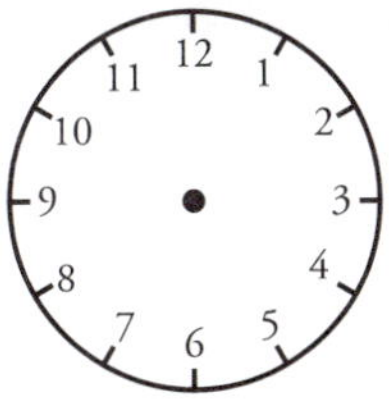   5 hours 23 minutes later

2.

2:44 starting time 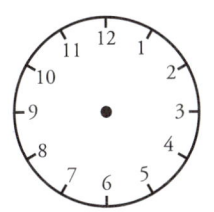   9 hours 12 minutes later

C Solve.

1.

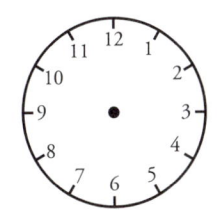

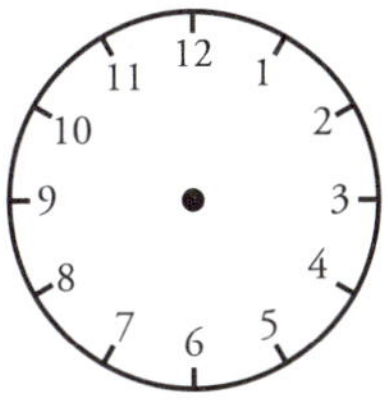

If it is 2:30 P.M. in New York, what time is it in Los Angeles, which is 3 hours earlier?

2.

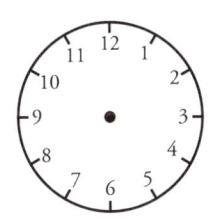

Morning circle started at 8:55 and lasted for 19 minutes. When did it end?

3.

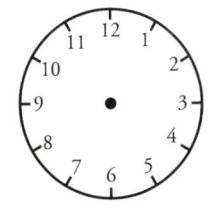

In a relay race, Tom ran the course in 20 seconds and gave the baton to Kyle, who ran it in 20 seconds. Leah took the final leg and ran it in 17 seconds. How many seconds did it take their team to run their part of the race?

Strategy: Draw a Clock Face   59

LESSON 4

STRATEGY: DRAW A MAP

1. Gather Information
2. Plan
3. Calculate
4. Check

On a car trip, Pete's family traveled at 55 miles an hour:

- east for two hours on Highway 80

    55 miles per hour x 2 hours = 110 miles

- north for one hour on Highway 10

    55 miles per hour x 1 hour = 55 miles

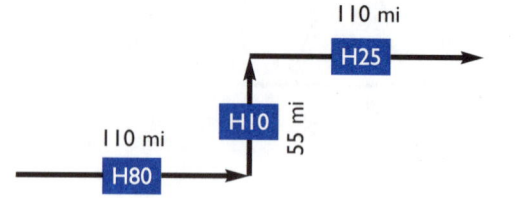

- east for two hours on Highway 25

    55 miles per hour x 2 hours = 110 miles

A Use the map to answer these questions.

1. After 5 hours, how far east did Pete's family go?

    Ⓐ 275 miles

    Ⓑ 165 miles

    Ⓒ 220 miles

    Ⓓ 110 miles

They traveled 110 miles east, then north (which doesn't count in the total), then 110 more miles east. The total distance traveled east is 220 miles so C is the right answer. Now try these on your own.

2. How many hours did they spend traveling east?

    Ⓐ 4 hours

    Ⓑ 3 hours

    Ⓒ 5 hours

    Ⓓ 2 hours

Unit 5: Time and Distance

3. What is the total number of miles Pete's family traveled during their 5-hour trip?

   Ⓐ 310 miles

   Ⓑ 265 miles

   Ⓒ 220 miles

   Ⓓ 275 miles

B Sketch and label a map to show the situation described.

1. Leon rowed a boat from the shore to the other side of the lake in 1 hour and 30 minutes. He rowed his boat at 5 miles per hour. How far did he travel?

2. Ms. Rodriguez lives in Greenville and rides her bike 5 miles to her school in Springdale on days she can. Last month, she rode 12 times. How many miles did she drive?

C Draw a map and solve. Show your work on another piece of paper.

1. How long did it take a car going 30 miles an hour to drive in a rectangle 60 miles east, then 30 miles south, 60 miles west and 30 miles north?

2. If it takes 3 hours in a plane flying east at 620 miles per hour to get there, how far is Mountainland from Sandy Island?

3. The schoolyard of Lincoln School is a square. Each side is 500 feet long. Sabrina can travel it by wheelchair at 100 feet per minute. How long would it take her to go all the way around the schoolyard?

Strategy: Draw a Map

LESSON 5

CHOOSE A STRATEGY

1. Gather Information
2. Plan
3. Calculate
4. Check

Here are some strategies that can help you solve time and distance problems.

Strategies
- Use a clock face
- Use the formula
  Rate x Time = Distance Traveled
- Make a drawing or map
- Use a time/distance graph
- Use a table

Here's an example of how you might use a table to solve a problem.

It takes Deedee and her older brother 20 minutes to walk to school, going at a speed of 3 miles per hour. They walk exactly one mile. If they leave at 8:00, how far have they gone by 8:15?

Miles	Minutes
3	60
2	40
1	20
15/20 = 3/4	15

A Choose the letter of the best answer.

1. A clock face could help you solve this problem by showing—

    Ⓐ how fast they were walking

    Ⓑ the distance they traveled

    Ⓒ how many minutes have gone by since 8:00

    Ⓓ what time it was when they got to school

    Clock faces can show how much time Deedee and her brother have been walking. The correct answer is C. Now try these on your own.

2. To make a time/distance graph of this problem you need—

    Ⓐ the time and distance traveled at different points of the trip

    Ⓑ the distance traveled

    Ⓒ the speed

    Ⓓ the starting time and the distance traveled

62   Unit 5: Time and Distance

3. Which solution to the problem correctly uses the formula *rate x time = distance traveled*?

   Ⓐ 3 miles per hour x 20 minutes = 60 miles

   Ⓑ 3 miles per hour x 1/3 hour = 1 mile

   Ⓒ 1 mile x 20 minutes = 20 miles per hour

   Ⓓ 3 miles per hour x 1 mile = 3 minutes

B **Write a few words to tell which strategy you would use to solve these problems, and tell why.**

1. On Kaylee's bus route, the driver goes east for 2 miles, south for 1 mile, and west for 3 miles. How far west does her bus travel?

2. Because of the time changes across the country, passengers on a plane going from Texas to New York have to set their watches forward 2 hours. If they start out at 1:00 on a 2 hour flight, what time should watches say when they get to New York?

C **Choose a strategy and solve. Show your work on another piece of paper.**

1. The space shuttle travels at 17 miles per second. After 6 seconds, it travels 102 miles. After 10 seconds, it travels 170 miles. How many miles will it travel after 5 seconds? 7 seconds? 8 seconds? 9 seconds?

2. Nico's school starts at 8:30 and ends at 3:05. How many hours does his school day last?

3. On Field Day, Ms. Makunda's class held a human wheelbarrow race down the length of a football field. If Tony and Hafiz could go the length of the 100 yard field in 2 minutes and 30 seconds, how long did it take them to go 10 yards?

LESSON 6

MULTI-STEP STORY PROBLEMS

1. Gather Information
2. Plan
3. Calculate
4. Check

Word problems about time and distance usually take more than one step to solve.

Sometimes you will add, subtract, multiply, or divide numbers first before you can solve the other parts of the problem.

Here's an example:

Captain Blasto got in the SuperCar at noon and drove at full speed. He arrived at the South Pole at 2:00 P.M., a trip of 3,200 miles. How fast did he drive?

Step 1: How long did Captain Blasto drive?

12:00 to 2:00 is 2 hours

Step 2: How fast did he drive?

3,200 in 2 hours

1,600 in 1 hour

So, he drove at 1,600 miles per hour.

A Identify the next step in solving the problem.

1. The movie "Kids Rule!" starts at 1:20 and ends at 3:00. How much longer is the movie "The Three Bears" which starts at 1:05 and ends at 3:10?

   Step 1: Find out how much time it is from 1:20 to 3:00, and 1:05 to 3:10

   Step 2: Subtract:

   Ⓐ 3:10 − 1:20
   Ⓒ the length of "Kids Rule!" from "Three Bears"

   Ⓑ 1:20 − 3:10
   Ⓓ the length of "The Three Bears" from "Kids Rule"

   In Step 1, we found the lengths of the movies. In Step 2, we subtract the shorter from the longer. D is the correct answer. Now try this on your own.

2. Lunch at Emilio's school goes from 11:45–12:15. Recess is from 2:15–2:45. How much time does Emilio spend at lunch and recess combined?

   Step 1: Find how many minutes lunch is, and how many minutes recess is.

   Step 2: Add:

   Ⓐ 11:45 and 2:15
   Ⓒ the times of lunch and recess together

   Ⓑ 11:45 and 2:45
   Ⓓ the ending time of recess from the starting time of lunch

64 Unit 5: Time and Distance

B List the steps in solving these multi-step problems.

1. Sally and her friends were walking to the game. They got 1 mile into their trip when they had to go home because they forgot their tickets. After the game, they walked two-and-a-half miles back. How far did they walk going to and from the game?

2. Every time Janel cut the lawn, she did it 3 minutes faster. The first time she started at 2:00 and finished at 2:45. How long did it take her on the fifth time?

C Solve. Show your work on another piece of paper.

1. Cheng-Hue's hair is 12 inches long. Victoria's hair is 3 inches shorter. Kirstin's hair is 2 inches longer than Victoria's. How long is Kirstin's hair?

2. It takes Edwardo 20 minutes to get ready for school. It takes Chris 3 minutes longer than Edwardo. It takes Lana 2 minutes less than Chris. Who takes longer to get ready, Lana or Edwardo?

3. August has 31 days, September has 30 and October has 31. November has 30 days, December has 31 and January has 31. Which 3 months have more days, August-October or November-January?

Multi-step Story Problems 65

UNIT 6 LESSON 1

SKILL TUNE-UP: MONEY AND PERCENTS

A Choose the correct answer.

1. $ 5.35
    × 9

    Ⓐ $29.08
    Ⓑ $28.07
    Ⓒ $4.81
    Ⓓ $48.15

2. 12 × $2.40 =

    Ⓐ $24.00
    Ⓑ $28.80
    Ⓒ $4.84
    Ⓓ $48.40

3. $1.98
    × 25

    Ⓐ $50.00
    Ⓑ $48.50
    Ⓒ $49.50
    Ⓓ $19.80

4. $ 0.50
    × 31

    Ⓐ $15.50
    Ⓑ $1.55
    Ⓒ $0.15
    Ⓓ $3.10

5. What is 50 percent of 90?

    Ⓐ 50
    Ⓑ 45
    Ⓒ 5.0
    Ⓓ 4500

6. What is 10 percent of $3.00?

    Ⓐ $30.00
    Ⓑ $0.30
    Ⓒ $3.30
    Ⓓ $1.30

B Write a number sentence you could use to solve the problem.

1. Chris wanted to buy a baseball mitt that cost $30. He also wanted to buy a bat for $14. How much will they cost in all?

Unit 6: Money and Percents

2. Mr. Greene's class bought 3 pizzas for a class picnic. They each cost $6.75. How much did they spend on their pizzas?

C Solve. Show your work.

1. How much would it cost to buy a $0.79 ruler for each student in the class if there are 27 students?

2. There are 44 students in the third grade. 25 percent of them are piano students. How many third graders are piano students?

3. When Ms. Hogan ordered $129 worth of art supplies for her class, she got a 30 percent discount. How much did these art supplies cost?

LESSON 2

STRATEGY: MAKE A TALLY CHART

1. Gather Information
2. Plan
3. Calculate
4. Check

A tally chart can help you keep track of things as they happen.

Sarah and Lizzy were selling refreshments at Lizzy's sister's softball game. They made one mark on their tally chart whenever they sold an item. This let them keep track of how much money they should have made. To calculate the total, they multiplied the price of the item times the number of tally marks.

Ice Cream Sandwiches $0.75	Cherry Ice $0.50	Lemon Ice $0.50	Iced Tea $0.95	Lemonade $0.85	Orangeade $0.85
$0.75 x 3	$0.50 x 4	$0.50 x 7	$0.95 x 9	$0.85 x 4	$0.85 x 5

A Use the tally chart to help you choose the correct answer.

1. How much money did they take in from selling ice cream sandwiches?
   - Ⓐ $2.25
   - Ⓑ $3.00
   - Ⓒ $.75
   - Ⓓ $1.50

There are 3 tally marks under the ice cream sandwich column. Each sandwich cost $0.75, so 3 x $0.75= $2.25. A is the correct answer. Now try these on your own.

2. How much money did they collect by selling iced tea?
   - Ⓐ $9.50
   - Ⓑ $8.55
   - Ⓒ $9.95
   - Ⓓ $8.10

3. How much money did they take in altogether?
   - Ⓐ $22.99
   - Ⓑ $24.89
   - Ⓒ $22.95
   - Ⓓ $23.95

Unit 6: Money and Percents

B Create a tally chart to show the information.

1. At the video store, new releases are $3.99. Older movies are $3.00. Video games are $2.00. In 1 hour, the store rented 3 new movies, 5 older movies, and 4 video games.

New Releases $3.99	Older Movies $3.00	Video Games $2.00

2. Edward started a pet sitting business over winter break. He charged 3 dollars a day to take care of dogs, 2.50 to take care of cats, and 2 dollars to feed fish. He had 2 dogs, 2 cats, and 1 fish tank on his customer list.

Dogs $3.00	Cats $2.50	Fish $2.00

C Use a tally chart to help you solve the problem. Show your work on another piece of paper.

Here is the top half of the receipt from Rita's trip to the store.

1. How much did she spend on sandwich bags?

2. How much did she spend on everything?

3. If she had a discount card that let her take 10 percent off, how much would these items cost?

```
SAV-O-RAMA GROCERY
5/15/04
2:45 P.M.
Cashier: Joyce

peanut butter      $2.49
peanut butter      $2.49
sandwich bags      $1.19
sandwich bags      $1.19
sandwich bags      $1.19
fruit jelly        $1.79
fruit jelly        $1.79
fruit jelly        $1.79
bakery (bread)     $1.65
bakery (bread)     $1.65
```

Strategy: Make a Tally Chart 69

1. Gather Information
2. Plan
3. Calculate
4. Check

L E S S O N 3

STRATEGY: USE PERCENTAGES OR PROPORTIONS

At Luke's school 45 out of every 100 students brought their lunch on Tuesday.

Students Who Brought Lunch From Home on Tuesday

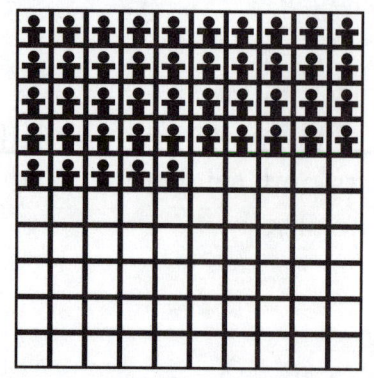

45/100 of the students are shaded.   45/100 = 45%

If there are exactly 200 students at school, how many brought their lunch on Tuesday?

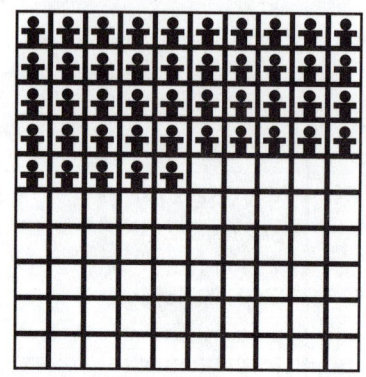

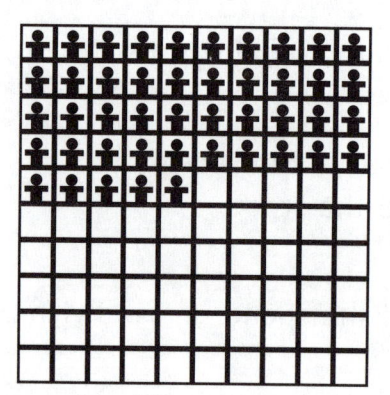

Here's one solution:

There are 200 students in all. 45 of the first 100 + 45 of the second 100 = 90 of the 200 students

Here's another solution:

45/100 = 45%   45% = 0.45   0.45 x 200 = 90 students

Here's a third solution:

45/100 = x/200   45/100 x 200 = x/200 x 200   9000/100 = x

x = 90 students

Unit 6: Money and Percents

A Choose the correct answer.

1. If there were 500 students at Luke's school, the number who brought their lunches from home on Tuesday would be—

   Ⓐ 250   Ⓑ 225   Ⓒ 90   Ⓓ 45

   Multiplying 0.45 x 500 gives us 225. The correct answer is B. Now try these on your own.

2. If 55 out of every 100 students brought their lunch on Tuesday, how many students would that be in all? Remember, Luke's school has 200 students.

   Ⓐ 55   Ⓑ 165   Ⓒ 110   Ⓓ 220

B Fill in the correct number of boxes to represent the amount described.

1. Shade 24% of these squares.

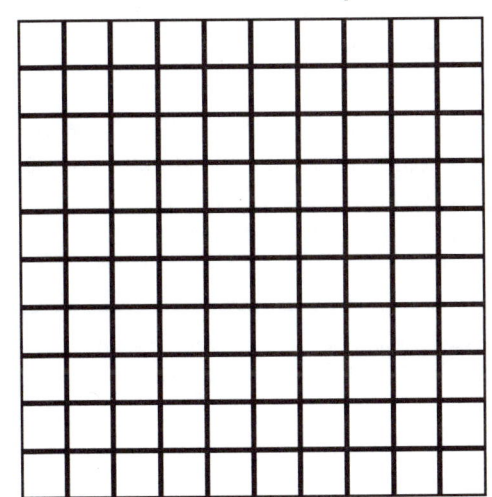

2. Shade 24% of these squares

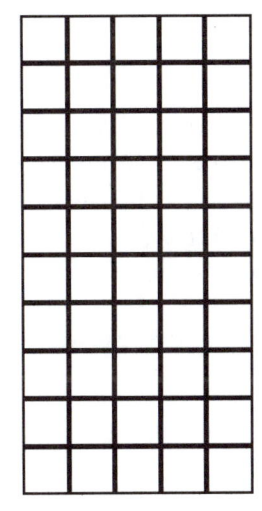

C Solve. Show your work.

1. 30 percent of the books in Shawana's school library are picture books. There are 1,660 books. How many are picture books?

2. 2 out of every 3 spaces in the parking lot were full. If there are 66 spaces, how many were full?

Strategy: Use Percentages or Proportions

LESSON 4

CHOOSE A STRATEGY

1 Gather Information
2 Plan
3 Calculate
4 Check

Here are some strategies you may find helpful when solving word problems about money and percents.

Strategies	Problem	How to use the strategy
Make a Tally Chart	Harry's family ordered three Superburgers for $1.55 each, and 2 medium-sized beverages for 95 cents each. How much did they spend?	Make a chart with the different items and put tally marks for each one ordered. Multiply the price times the number of marks in each column.
Draw a Picture		Draw pictures of 3 burgers and 2 beverage cups. Label each one with its price. Add the prices together to get the total.
Use Manipulatives (such as "school money")		Count out 3 piles of coins and bills that equal $1.55. Count out 2 piles of coins that equal 95 cents. Count all this money to get the total.
Use Proportions	Ms. Raines's class planned to give 20 percent of the money they collected at their class fundraiser to a food kitchen. If they collected $118, how much did they give to the food kitchen?	10 percent of $118 is $11.80. 20 percent is double this amount or $11.80 + $11.80.

A **Choose the best strategy to solve the problems below.**

1. Janel had 32 quarters, 12 dimes, and 70 nickels in her piggy bank. What was the total value of these coins?

   Ⓐ make a tally chart   Ⓒ draw a picture

   Ⓑ use proportions   Ⓓ any of these methods would work

Making a tally chart would help you figure out the value of each of the three kinds of coins she has. The correct answer is A. Now try this on your own.

72 Unit 6: Money and Percents

2. 68% of the students at Williams Elementary are in families with at least 1 brother and 1 sister. There are 510 students in the school. How many have at least 1 brother and 1 sister?

   Ⓐ draw a picture
   Ⓑ make a tally chart
   Ⓒ use percentages or proportions
   Ⓓ work backwards

B Solve. Make a tally chart if it helps. Show your work.

1. Here's what Joe found on the beach: 14 conch shells, 31 clam shells, 19 whelk shells, and 9 scallop shells. How many shells did he find?

2. Ms. Hurston's class made a very big cake. They divided it into 4 pieces and sold each piece in a raffle. The top left piece sold for $8.55. The top right piece sold for $6.25. The bottom left piece sold for $5.20. The bottom right piece sold for $7.75. How much money did they collect?

3. Teenagers Sal and John worked at summer jobs and each made $800. They each gave 20 percent of the money they made to their parents for a college fund. How much did each save for his college fund?

4. What is the value of 15 half-dollars, 23 quarters, 77 dimes, 34 nickels, and 28 pennies?

Choose a Strategy   73

LESSON 5

MULTI-STEP STORY PROBLEMS

1. Gather Information
2. Plan
3. Calculate
4. Check

Most math problems involving both money and percents are multi-step problems. The first step is usually to add, subtract, multiply, or divide amounts of money. The second step is often to find a percentage of that total.

Here's an example:

On a trip to the store, Kelvin spent 15 cents on jaw breakers, 15 cents on licorice, and 20 cents on jelly beans and 10 cents on chocolate kisses. What percentage of this money did he spend on licorice?

First step: What was the amount of money spent?

15 cents + 15 cents + 20 cents + 10 cents = 60 cents

Next step: What percentage of this 60 cents is 15 cents?

? % × 60 cents = 15 cents

0.25 = 15 ÷ 60

0.25 = 25 percent

A What would be a good next step to solve these problems?

1. The 4th grade bake sale raised $23 in Ms. Gold's class, $35 in Ms. Elder's class, and $42 in Ms. Alper's class. The classes gave 15 percent of their sales to the library. How much did they give to the library?

   First step: Add the totals raised in each class.
   Next step:

   Ⓐ subtract 15 from the total
   Ⓑ add 15 to the total
   Ⓒ multiply the total by 0.15
   Ⓓ multiply the total by 1500

   The correct answer is C. Now try these on your own.

2. Jan bought a stereo for $89.95 and speaker wire for $17.95. If there was a 5 percent sales tax, how much did she pay?

   First step: Add the cost of the stereo and speaker wire.
   Next step:

   Ⓐ multiply the total by 5
   Ⓑ multiply the total by 0.05
   Ⓒ add $5.05 to the total
   Ⓓ subtract $5 from the total

Unit 6: Money and Percents

B Describe your steps for solving the problem.

1. Mr. Duke bought 40 composition books for his class at 85 cents each. The store gave him a 10 percent discount because it was for a school. How much did he pay?

    First step:

    Next step:

2. Four coaches spent $360 on a case of baseballs. One of the coaches paid half of the cost. The others split the rest of the cost evenly. How much did each pay?

    First step:

    Next step:

C Solve. Show your work on another piece of paper.

1. Matt bought two packs of guitar strings for $6.70 each and had a coupon for 50 percent off on his next pack. He bought his next pack right away. How much did he pay for the 3 packs of guitar strings?

2. A penny drive raised $215 on day 1 and $121 on day 2. What percentage of the total was raised on day 1?

UNIT 7   LESSON 1

SKILL TUNE-UP: ADDING AND SUBTRACTING FRACTIONS

1. Gather Information
2. Plan
3. Calculate
4. Check

A What part of the shape is NOT shaded?

1. Shaded part: $\frac{3}{8}$; non-shaded part: ___

   Ⓐ $\frac{3}{8}$ Ⓑ $\frac{5}{8}$ Ⓒ $\frac{5}{5}$ Ⓓ $\frac{5}{3}$

There are 8 parts to the circle. Three are shaded, 5 are not. 5/8 of the circle is not shaded. B is the correct answer. Now try these on your own.

2. Shaded part: $\frac{4}{6}$; non-shaded part: ___

   Ⓐ $\frac{6}{2}$ Ⓑ $\frac{3}{6}$ Ⓒ $\frac{2}{2}$ Ⓓ $\frac{2}{6}$

3. Shaded part: $\frac{1}{4}$; non-shaded part: ___

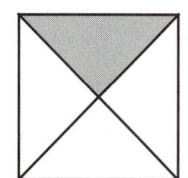

   Ⓐ $\frac{1}{4}$ Ⓑ $\frac{2}{4}$ Ⓒ $\frac{1}{3}$ Ⓓ $\frac{3}{4}$

4. Shaded part: $\frac{5}{6}$; non-shaded part: ___

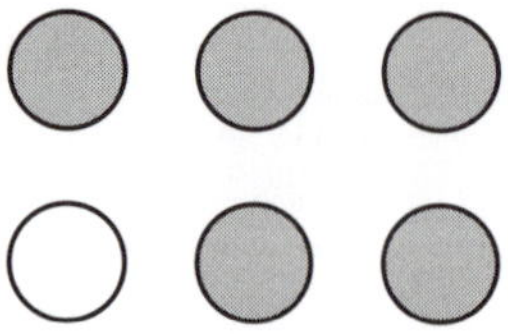

   Ⓐ $\frac{5}{6}$ Ⓑ $\frac{2}{6}$ Ⓒ $\frac{1}{6}$ Ⓓ $\frac{1}{4}$

5. Shaded part: $\frac{1}{3}$; non-shaded part: ___

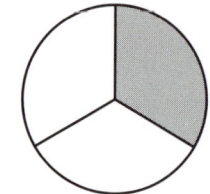

   Ⓐ $\frac{1}{2}$ Ⓑ $\frac{2}{2}$ Ⓒ $\frac{3}{4}$ Ⓓ $\frac{2}{3}$

6. Shaded part: $\frac{2}{3}$; non-shaded part: ___

   Ⓐ $\frac{1}{3}$ Ⓑ $\frac{3}{3}$ Ⓒ $\frac{3}{6}$ Ⓓ $\frac{1}{8}$

Unit 7: Adding and Subtracting Fractions

B Solve.

1. $\frac{2}{7} + \frac{3}{7} =$

2. $\frac{1}{9} + \frac{1}{9} =$

3. $\frac{5}{6} - \frac{1}{6} =$

4. $\frac{3}{4} - \frac{1}{4} =$

C Write a mixed number for each picture.

1. ____

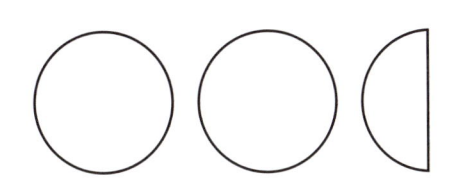

4. ____

2. ____

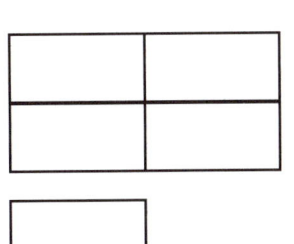

3. ____

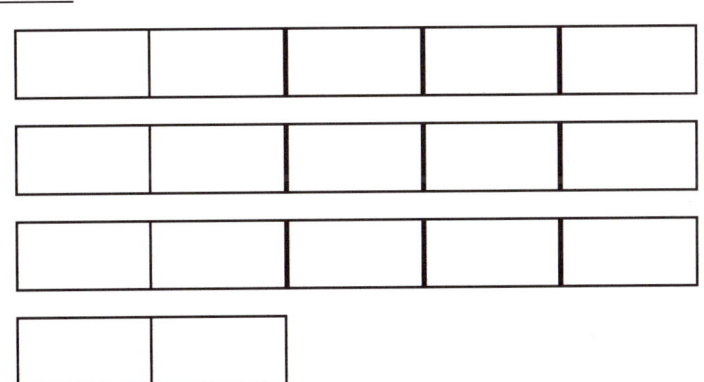

Skill Tune-up: Adding and Subtracting Fractions

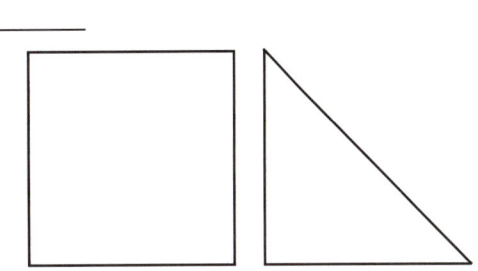

LESSON 2

STRATEGY: MAKE A CIRCLE GRAPH

A circle graph can help you see how equal fractions fit together to make a whole.

The circle below is cut into four equal pieces. Each piece is $\frac{1}{4}$ of the circle.
$\frac{1}{4} + \frac{1}{4} + \frac{1}{4} + \frac{1}{4} = \frac{4}{4}$.

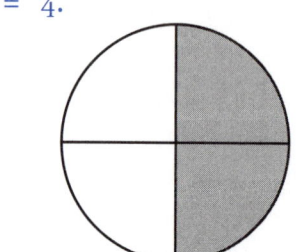

Notice that two of the sections are shaded.

$\frac{1}{4} + \frac{1}{4} = \frac{2}{4}$

$\frac{2}{4}$ is another way to say that the shaded area is 2 pieces out of 4.

You might also notice that $\frac{1}{2}$ of the circle is shaded. $\frac{1}{2}$ is another way to say $\frac{2}{4}$.

$\frac{2}{4} = \frac{1}{2}$

A Choose the best answer.

1. What part is shaded?

   Ⓐ $\frac{1}{2}$ Ⓑ $\frac{2}{3}$ Ⓒ $\frac{3}{4}$ Ⓓ $\frac{1}{3}$

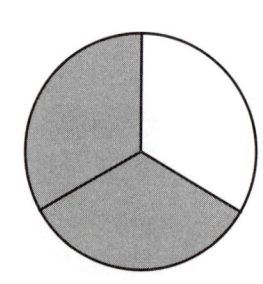

The pie graph has 3 pieces and one of them is shaded. 1/3 is shaded, so D is the correct answer. Now do these yourself.

2. What part is shaded?

   Ⓐ $\frac{1}{2}$ Ⓑ $\frac{6}{8}$ Ⓒ $\frac{2}{8}$ Ⓓ $\frac{2}{3}$

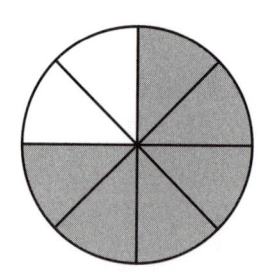

Unit 7: Adding and Subtracting Fractions

3. Which of these is $\frac{7}{10}$ shaded?

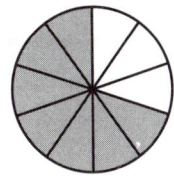

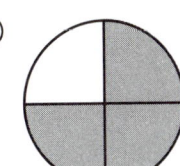

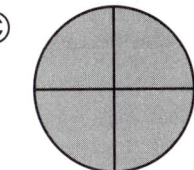

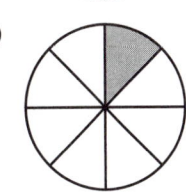

Ⓐ   Ⓒ

Ⓑ   Ⓓ

B Make a circle graph to help solve the problem.

1. On Sakshi's birthday, her mom made a pumpkin pie for her and her brothers and sisters. They cut it into six pieces and ate 4. How much of the pie was left?

2. Seamus made a circle graph to show how much money his class had raised compared to their goal for the penny drive. Their goal was $8.00, but they had raised $3.00. What did his chart look like?

3. Mark, his brother, and his dad were at the mall at noon. Mark's dad told the boys they could go on their own for 2/3 of an hour. How many minutes were they on their own?

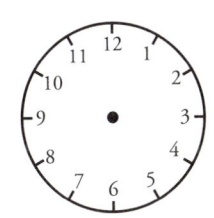

4. Tom ordered a pizza with 12 slices. 4 had broccoli. What fraction of the pizza did not have broccoli?

Strategy: Make a Circle Graph   79

LESSON 3

STRATEGY: FINDING EQUIVALENT FRACTIONS

1. Gather Information
2. Plan
3. Calculate
4. Check

The fraction 2/4 means the same thing as 1/2. They are equivalent fractions. We can see this when we divide a chocolate bar that has four equal sections.

We can break it into 4 parts

| $\frac{1}{4}$ | $\frac{1}{4}$ | $\frac{1}{4}$ | $\frac{1}{4}$ |

$\frac{1}{4} + \frac{1}{4} = \frac{2}{4}$

or 2 parts

| $\frac{1}{2}$ | |

$\frac{1}{2} = \frac{1}{2}$ $\frac{2}{4} = \frac{1}{2}$

One way is to divide something into equal fraction parts and then compare how many of one sized part is the same as another sized part. This is what we did with the chocolate bar.

Another way to find equivalent fractions is to multiply or divide both the numerator and denominator (the top and bottom numbers of a fraction) by the same number.

$\frac{1}{2} \times \frac{2}{2} = \frac{2}{4}$

A Choose the best answer.

1. Which is another way of saying $\frac{3}{4}$?
   Ⓐ $\frac{2}{3}$ Ⓑ $\frac{6}{9}$ Ⓒ $\frac{9}{12}$ Ⓓ $\frac{4}{5}$

   Because $\frac{3}{4} \times \frac{3}{3} = \frac{9}{12}$, C is the correct answer. Now try these on your own.

2. Which fraction is equivalent to $\frac{2}{3}$?
   Ⓐ $\frac{10}{15}$ Ⓑ $\frac{9}{16}$ Ⓒ $\frac{3}{4}$ Ⓓ $\frac{3}{5}$

3. Another way to say $\frac{1}{2}$ is—
   Ⓐ $\frac{2}{3}$ Ⓑ $\frac{4}{6}$ Ⓒ $\frac{6}{12}$ Ⓓ $\frac{3}{4}$

80   Unit 7: Adding and Subtracting Fractions

B Solve these problems. Show your work.

1. Vlad cut $\frac{1}{3}$ of the lawn on Saturday and another $\frac{1}{6}$ on Sunday. His sister did the rest. How much did Vlad cut?

2. $\frac{3}{12}$ of the children at Martin's school have cats and $\frac{1}{6}$ have fish. What fraction have either a cat or fish?

3. Lonnie made a pie. His brother ate $\frac{1}{4}$ of it. His sisters ate $\frac{1}{2}$ of it. How much was left?

4. Andy divided a large square into 100 small squares. He shaded 40 of the small squares. Mandy divided a square of the same size into 10 rectangles. She shaded 4 parts.

Andy's Square

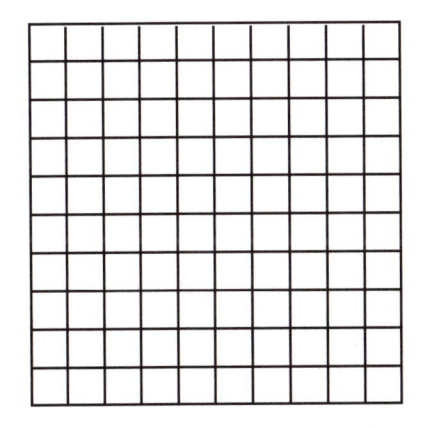

Mandy's Square

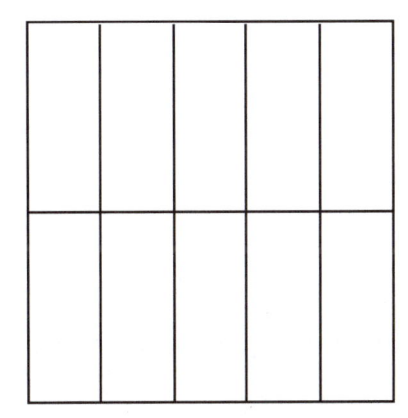

Andy says that he shaded more of his square than Mandy did of hers. Mandy says she shaded the same amount as Andy. Who is correct? Explain your answer.

Strategy: Finding Equivalent Fractions

LESSON 4

CHOOSE A STRATEGY

1. Gather Information
2. Plan
3. Calculate
4. Check

Here are some strategies for solving problems that ask you to add or subtract fractions.

Strategies	How to use the strategy
Make a circle graph	Draw circles and divide them into equal parts. Compare or add the sizes of the parts.
Shade regions	Draw a shape and shade an amount of it equal to a fraction.
Use sets	Divide a group of objects into equal parts. Mark or set aside a number of equal groups.
Use fraction strips	Compare the lengths of equal bars divided in different ways.
Make an equivalent fraction	Multiply the numerator and denominator by the same number.

A Choose the best strategy to solve the problems below.

1. Ms. Myers' class has 14 students. 7 are boys. What fraction of the class is boys?

   Ⓐ use sets
   Ⓑ shade regions
   Ⓒ make a pie graph
   Ⓓ make equivalent fractions

   By counting out 7 students of the 14 in the class, we can see that 7 students is $\frac{1}{2}$ of the class. A is the correct answer. Now try these on your own.

2. A pizza was cut into 10 equal pieces. 3 were eaten. What fraction was left?

   Ⓐ make equivalent fractions
   Ⓑ shade regions
   Ⓒ make a circle graph
   Ⓓ use fraction strips

82   Unit 7: Adding and Subtracting Fractions

3. Delia wanted to know which fraction was bigger, $\frac{22}{40}$ or $\frac{30}{50}$. How could she be sure?

   Ⓐ make a circle graph   Ⓒ use sets
   Ⓑ make equivalent fractions   Ⓓ both B and C

B Explain how you would use the given strategy to solve the problem. The first one is done for you.

1. Use Sets

   $\frac{4}{5}$ of the 3rd graders chose outdoor recess on a cold and drizzly day. 50 children were in school. How many chose outdoor recess?

   Answer: Draw 50 marks. Divide them into 5 equal groups. Circle 4 of the groups. Count the number of marks.

2. Shade Regions

   How can you fold a rectangular piece of paper to solve the equation $\frac{8}{8} - \frac{3}{8} = ?$

C Solve. Show or explain how you got your answer.

1. Sydney cut a pie into 16 equal pieces and ate 3. Joe cut an identical pie into 8 pieces and ate 2. Who had more pie?

2. A deck has 52 cards. What fraction, with a denominator of 52, is the same as $\frac{1}{4}$ of a deck?

3. Dan and Tom were collecting pennies. Dan had $\frac{1}{2}$ of a dollar in pennies. Tom had $\frac{1}{20}$ of a dollar. How many pennies did they have altogether?

Choose a Strategy

LESSON 5

MULTI-STEP STORY PROBLEMS

1. Gather Information
2. Plan
3. Calculate
4. Check

Rick's family had two loaves of bread. They ate $\frac{5}{12}$ of a loaf of bread on Monday. On Tuesday, they ate $\frac{1}{3}$ of a loaf. On Wednesday, they ate another $\frac{3}{4}$ of a loaf. Was there any bread left? If so, how much?

Step 1: Find equivalent fractions.

Monday $\quad \frac{5}{12} = \frac{5}{12}$

Tuesday $\quad \frac{1}{3} \times \frac{4}{4} = \frac{4}{12}$

Wednesday $\quad \frac{3}{4} \times \frac{3}{3} = \frac{9}{12}$

Step 2: Add or subtract.

$\frac{5}{12} + \frac{4}{12} + \frac{9}{12} = \frac{18}{12}$ of a loaf

Step 3: Simplify. Change improper fractions to mixed numbers.

$18 \div 12 = 1$ remainder 6

$1 \frac{6}{12}$

Step 4: Make other calculations.

$2 - 1\frac{6}{12} = \frac{6}{12}$, or $\frac{1}{2}$ of a loaf.

A Choose the best answer.

1. Another way to say $\frac{12}{8}$ is —

   Ⓐ 1/2/8

   Ⓑ $12\frac{8}{12}$

   Ⓒ $8\frac{4}{12}$

   Ⓓ $1\frac{1}{2}$

When we divide $12 \div 8$ we get 1 remainder 4. We write this as a fraction as $1\frac{4}{8}$, or $1\frac{1}{2}$. D is the correct answer. Now try these on your own.

2. The fraction $\frac{16}{5}$ can also be written as—

   Ⓐ $8\frac{2}{5}$

   Ⓑ $\frac{5}{16}$

   Ⓒ $3\frac{1}{5}$

   Ⓓ $\frac{31}{5}$

3. The mixed number $3\frac{1}{3}$ can also be written as—

   Ⓐ $\frac{10}{3}$

   Ⓑ $\frac{3}{10}$

   Ⓒ $\frac{9}{3}$

   Ⓓ $\frac{3}{9}$

Unit 7: Adding and Subtracting Fractions

B Write a few words or sentences to describe your plan for solving the problem.

1. Nadia had a dollar. $\frac{1}{2}$ of it was in quarters. $\frac{1}{5}$ of it was in dimes. The rest was in nickels. What fraction of it was in nickels?

2. April has 30 days. This year 10 of the days are weekend days. What fraction of April days are weekdays?

C Solve each problem. Show your work.

1. There are 3 tennis balls in a package. Ms. Albano has 20 packages in the gym. Ms. Pennibaker's class used 10 tennis balls. What fraction of the balls were still unused?

2. Mr. Gerber's class collected coins in their savings jar. Of the coins, $\frac{1}{5}$ were quarters and $\frac{2}{5}$ were dimes. The rest were pennies. What fraction of this money were pennies?

3. Elisha drank $\frac{2}{3}$ of a quart of water on a hike. Shand drank $\frac{3}{5}$ of a quart. How many quarts did they drink?

Multi-step Story Problems

UNIT 8 LESSON 1

SKILL TUNE-UP: GEOMETRY AND MEASUREMENT

A Choose the letter that matches the figure shown.

1. 
   Ⓐ square   Ⓑ rectangle   Ⓒ trapezoid   Ⓓ rhombus

2. 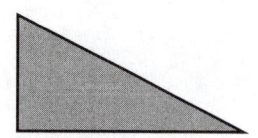
   Ⓐ right angle   Ⓒ isosceles triangle
   Ⓑ equilateral triangle   Ⓓ right triangle

3. 
   Ⓐ hexagon   Ⓑ rectangle   Ⓒ parallelogram   Ⓓ trapezoid

4.

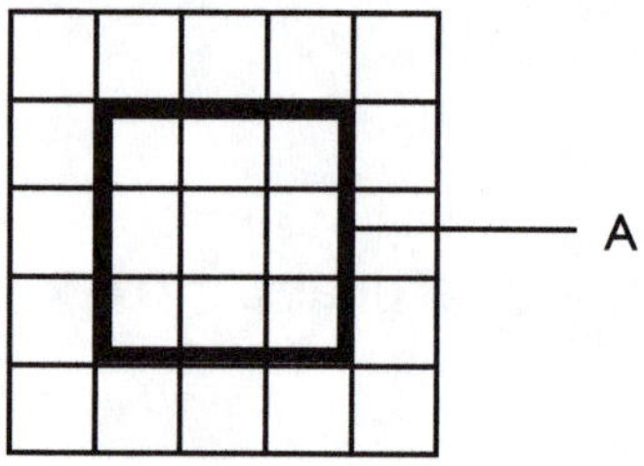

   If each small square is one square foot in area, what is the area of square A?
   Ⓐ 3 square feet   Ⓒ 25 square feet
   Ⓑ 6 square feet   Ⓓ 9 square feet

86   Unit 8: Geometry and Measurement

5. What is the perimeter of square A?

   Ⓐ 3 feet Ⓑ 6 feet Ⓒ 25 feet Ⓓ 9 feet

6. What is the perimeter of rectangle D?

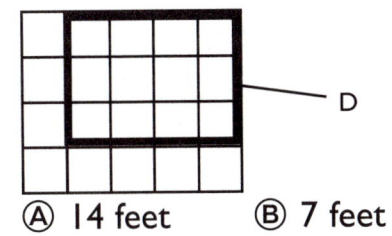

   Ⓐ 14 feet Ⓑ 7 feet Ⓒ 12 feet Ⓓ 16 feet

B Draw these figures with a ruler.

1. A square with each side measuring 2 inches.

2. A rectangle with one side of 2 inches and one side of 1 inch.

C Solve. Show your work.

1. What is the perimeter of this rectangle?

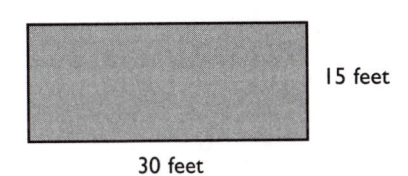

   15 feet
   30 feet

2. If we divide this square in half so that it makes two triangles, what is the area of each of the triangles?

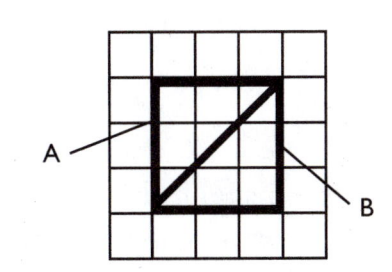

Skill Tune-up: Geometry and Measurement   87

LESSON 2

STRATEGY: DRAW A DIAGRAM

1. Gather Information
2. Plan
3. Calculate
4. Check

When carpenters build something, they often start by drawing a diagram. A diagram helps to see how things fit together, and how they relate to each other. A simple diagram might look like this.

5 foot board

| 10 inches | $2\frac{1}{2}$ feet | $2\frac{1}{2}$ feet | 10 inches |

This diagram tells us that the board is 5 feet wide and 10 inches high. We can also see that the board was sawed in half to make two smaller boards that are each $2\frac{1}{2}$ feet wide.

A diagram may tell us the shape of an object, the size, and the distance one part is from another.

Making or using a diagram can help you see or show important information that you need to solve a word problem.

A Use the diagrams to help you choose the correct answer.

6 inches

1. What is the perimeter of this hexagon that has sides that are all of equal length?

   Ⓐ 6 inches   Ⓑ 30 inches   Ⓒ 36 inches   Ⓓ 48 inches

   The diagram shows us that there are 6 sides and each side is 6". 6 x 6" = 36" so C is the correct choice. Now try these on your own.

88   Unit 8: Geometry and Measurement

15 feet

60 feet

2. What is the area of this rectangle?

   Ⓐ 900 square feet
   Ⓑ 900 feet
   Ⓒ 75 square feet
   Ⓓ 150 feet

3. Which has more area, diamond A or square B?

   Ⓐ diamond A
   Ⓑ square B
   Ⓒ A and B are equal
   Ⓓ it is not possible to tell

B Make a diagram to match the information given in these word problems.

1. The recess field at Ankor's school is a rectangle 400' x 100'. Make a drawing of it and label how long each side is.

2. Stu had a piece of chocolate that was cut into squares 1" x 1". It was 2 inches high and 5 inches long. Show what it looks like.

C Use a diagram to solve. Show your work on another sheet of paper.

1. The lunchroom at Jerry's school was a rectangle 70 feet wide and 60 feet long. Over the summer, the school put up dividers to make the dining area feel more cozy. The dividers cut the room into four equal parts. What did it look like and what was the area of these smaller sections?

Strategy: Draw a Diagram   89

LESSON 3

STRATEGY: USE A FORMULA

1. Gather Information
2. Plan
3. Calculate
4. Check

There are many times that you may want to know the perimeter or area of a regular shape such as a square, a rectangle, a circle, a triangle, or other geometric figure.

We need this information when we cut fabric, paint rooms, build shelves, or even put frosting on cakes.

Perimeter is the distance around the outside of a shape.

The formula for finding the perimeter of a 4-sided figure is

Side$_1$ + Side$_2$ + Side$_3$ + Side$_4$ = Perimeter

8 in. + 14 in. + 8 in. + 14 in. = 34 in.

Area is the amount of space inside the boundaries of a shape.

Area is measured in square inches. The formula for finding the area of a rectangle is:

Height x Width =

Side 1 x Side 2 =

8 in. x 14 in. = 112 square inches

side 1 = 8 in.
side 2 = 14 in.
side 3 = 8 in.
side 4 = 14 in.

A Choose the best answer.

1. Which of these equations would help you find the perimeter of rectangle Q?

   side 1 = 4 yds.
   side 2 = 9 yds.
   side 3 = 4 yds.
   side 4 = 9 yds.

   Ⓐ 9 yds. + 4 yds. = 13 yds.

   Ⓑ 9 yds. x 4 yds. x 9 yds. x 4 yds. = 1296 yds.

   Ⓒ 9 yds. + 4 yds. + 9 yds. + 4 yds. = 26 yds.

   Ⓓ 9 yds. + 4 yds. + 9 yds. = 22 yds.

The formula for finding the perimeter of a rectangle is Side$_1$ + Side$_2$ + Side$_3$ + Side$_4$ = Perimeter. Choice C is the correct answer. Now try these on your own.

2. Which of these equations would help you find the *area* of Rectangle Q?
   - Ⓐ 9 yds. + 4 yds. = 13 yds.
   - Ⓑ 9 yds. + 4 yds. + 9 yds. + 4 yds. = 26 yds.
   - Ⓒ 9 yds. x 4 yds. x 9 yds. x 4 yds. = 1296 yds.
   - Ⓓ 9 yds. x 4 yds. = 36 yds.

3. If all the sides of a rectangle were 4 yards long, what would its *perimeter* be?
   - Ⓐ 4 yds. + 4 yds. + 4 yds. + 4 yds. = 16 yds.
   - Ⓑ 4 yds. + 4 yds. = 8 yds.
   - Ⓒ 4 yds. + 4 yds. + 4 yds. = 12 yds.
   - Ⓓ 4 yds.

B **Imagine you have a rectangle that is 7 inches tall and 18 inches wide. Now answer these questions.**

1. What numbers would you add together to find the perimeter of the rectangle?

2. What numbers would you multiply together to find the area of the rectangle?

C **Solve. Show your work on another piece of paper.**

1. Martin helped his class build sets for a play. Part of the set was a fenced in rectangle that was 9 feet long and 12 feet wide. How many feet of cardboard did they need to make this fence?

2. If Tyronne's classroom is a rectangle 41 feet long and 20 feet wide, what is the area of his room in square feet?

Strategy: Use a Formula

LESSON 4

CHOOSE A STRATEGY

1. Gather Information
2. Plan
3. Calculate
4. Check

Strategies	When to use the strategy	How the strategy is used
Draw a diagram	Use when you need to understand what goes where and what the relationships are between the measurements.	What is the area of each part of a 9" x 12" foot office if you cut it in half? 12 ft. 9 ft.
Use a formula	Use when you can easily find the measurements you need to calculate the area or perimeter of a regular shape.	The four sides of a square are 12 feet long. What is its area? Width x Height = Area 12 feet x 12 feet = Area
Use proportions	Use when you already know the area or perimeter of a figure that is the same, only a certain number of times bigger or smaller than the original.	If the perimeter of a fence is 88 feet, what is the perimeter of a fence twice as big? Perimeter of a fence = 88 feet Perimeter of a fence twice as big = 88 feet x 2
Use a measuring tape	Use if you can measure a real thing.	What is the perimeter of your desk? Take a measuring tape and wrap it around the edge of your desk and see what the total distance around is.
Use graph paper	Use if you are trying to find the area or perimeter of something with a regular shape.	What is the area of your 30 x 22 living room? Draw a rectangle 30 squares long and 22 squares wide on graph paper. Count the number of squares inside the rectangle you drew.

A Choose the best strategy.

1. Tom knows that the area of the roof of his school bus is 100 square feet. What is the area of a school bus half as long?

   Ⓐ Use graph paper
   Ⓑ Use a measuring tape
   Ⓒ Use a formula
   Ⓓ Use proportions

   Using proportions, we can see that something half as big has half as much area. D is the correct answer. Now try these on your own.

92   Unit 8: Geometry and Measurement

2. Find the perimeter of a box in your classroom.

   Ⓐ Use graph paper Ⓒ Use a formula

   Ⓑ Use a measuring tape Ⓓ Use proportions

3. What is the area of this rectangle?

   6 feet

   17 feet

   Ⓐ Draw a diagram Ⓒ Use a formula

   Ⓑ Use a measuring tape Ⓓ Use proportions

B Explain how you would use the strategy listed to solve each problem.

1. How could you use a measuring tape to find the perimeter of a flower bed?

2. How could you use a diagram to find the perimeter of a rectangular yard with sides that are 33 feet and 20 feet?

C Solve. Show your work.

1. The foundation of a house is a rectangle 20 yards by 15 yards. What is the area of the foundation?

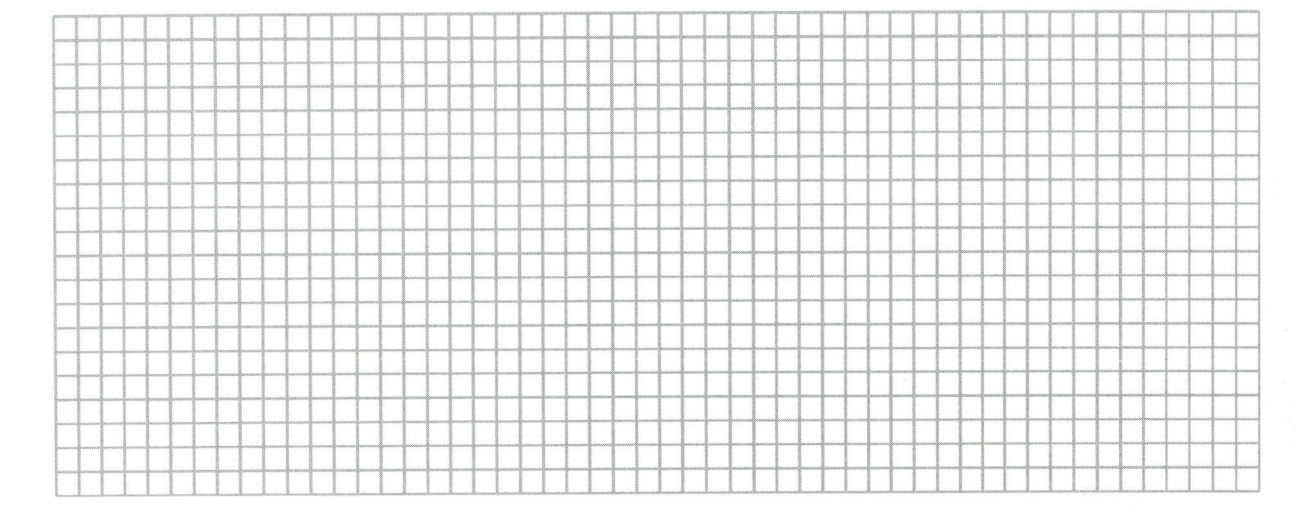

Choose a Strategy   93

LESSON 5

MULTI-STEP STORY PROBLEMS

1 Gather Information
2 Plan
3 Calculate
4 Check

In multi-step problems, you will be doing a series of calculations. It helps if you break the problem down step-by-step to help you develop a plan for coming up with your answer.

Joey is helping his mom by cutting the grass. The front lawn is a rectangle that is 30 feet by 70 feet. The back lawn is 70 feet by 100 feet. How many square feet of lawn would he cut if he did the front and back?

First: Calculate the area of the front lawn.

30 ft. x 70 ft. = 2100 square feet

Then: Calculate the area of the back lawn.

70 ft. x 100 ft. = 7000 square feet.

Finally: Add together the area of the front lawn and the back lawn.

2100 square feet + 7000 square feet = 9100 square feet

A Choose the best second step for these multi-step problems.

1. A climbing wall is 24 feet wide. It is half as tall as it is wide. What is the total area of the wall?

   First: Find out how tall the wall is.

   Then:

   Ⓐ add the height to the width

   Ⓑ multiply the width by 24

   Ⓒ add the width and height of the 4 sides

   Ⓓ multiply the height times the width

   When we know the height and the width, we can multiply them together to calculate the area. D is the correct answer. Now try these own your own.

Unit 8: Geometry and Measurement

2. How much more area does an 8' x 6' rug have than one that is 10' x 4'?

   First: Find the area of the 8' x 6' rug and the 10' x 4' rug by multiplying the width and height of each.

   Then:

   Ⓐ add the areas together

   Ⓑ subtract the area of the 10' x 4' from the area of the 8' x 6' rug

   Ⓒ subtract the perimeter of the 8' x 6' rug from the 10' x 4' rug

   Ⓓ add the areas of both rugs together and divide by 2

3. In the library, each shelf is 12 inches deep and 3 feet wide. If a book case has 4 shelves, how much total area is there to shelve books?

   First: Find the area of a shelf by multiplying 12 inches x 36 inches.

   Then:

   Ⓐ add 12 inches + 36 inches + 12 inches + 36 inches

   Ⓑ multiply that total by 4

   Ⓒ multiply 12 inches x 3

   Ⓓ add 3 + 12 + 4

B Describe the steps you would use to solve these multi-step problems.

1. Chris was making paper mats to go on top of the tables in the dining room. The tables are 8 feet long and 4 feet wide. There are 10 tables. How many square feet of paper will he need?

   First:
   Then:

2. Yvonne's school doubled the size of its garden from a rectangle 36' x 12' to one that is 36' x 24'. How much fence did her classmates have to add?

   First:
   Then:
   Finally:

Multi-step Story Problems   95